Principles of Verifiable RTL Design

Principles of Verifiable RTL Design

A functional coding style supporting

verification processes in Verilog

Lionel Bening and Harry Foster

Hewlett-Packard Company

Distributors for North, Central and South America:
Kluwer Academic Publishers
101 Philip Drive, Assinippi Park
Norwell, Massachusetts 02061 USA
Telephone (781) 871-6600
Fax (781) 871-6528
E-Mail <kluwer@wkap.com>

Distributors for all other countries:
Kluwer Academic Publishers Group
Distribution Centre
Post Office Box 322
3300 AH Dordrecht, THE NETHERLANDS
Telephone 31 78 6392 392
Fax 31 78 6546 474
E-Mail <orderdept@wkap.nl>

Electronic Services <http://www.wkap.nl>

Library of Congress Cataloging-in-Publication Data
Bening, Lionel, 1939-
  Principles of verifiable RTL design : a functional coding style supporting verification
processes in Verilog / Lionel Bening and Harry Foster.
    p. cm.
  Includes bibliographical references and index.
  ISBN 0-7923-7788-5 (alk. paper)
  1. Integrated circuits--Very large scale integration--Computer-aided design. 2.
Verilog (Computer hardware description language) 3. Electronic digital
computers--Computer-aided design. I. Foster, Harry, 1956- II. Title.

  TK7874.75. B47 2000
  621.39'2--dc21   00-021223

Printed on acid-free paper.

Printed in the United States of America

Dedicated to:

Ann, Laura and Donna

-Lionel

Roger, Elliott, Lance, Hannah and Jeanne

-Harry

Table of Contents

6 Verifiable RTL Style 123

PREFACE

The conception of a verifiable register transfer level (RTL) philosophy is a product of two factors: one, inherited seat-of-the-pants experiences during the course of large system design; the other, the sort of investigation which may be called "scientific." Our philosophy falls somewhere between the knowledge gained through experiences and the knowledge gained through scientific research. It corroborates on matters as to which definite knowledge has, so far, been ascertained; but like science, it appeals to reason rather than authority. Our philosophy consists of a fundamental set of principles, which when embraced, yield significant pay back during the process of verification.

The need for a verifiable RTL philosophy is justified by the complexity, density, and clock speeds of today's chips and systems, which continue to grow at exponential rates. This situation has raised the cost of design errors to a critical point--where, increasingly, the resources spent on the process of verification exceeds those spent on design.

Myriad books, manuals, and articles have addressed the issue of RTL Verilog style with an emphasis on synthesis-oriented policies. They explain how to write Verilog to wrest optimal gates from the synthesis process. Still other material presents the entire spectrum of Verilog constructs from the architectural specification to switch-level strengths. Yet, these works leave it to the readers to find their way to good practices for verification. Unfortunately, few guidelines govern the coding of RTL Verilog to achieve an optimum flow through the various functional and logical verification processes.

This vacuum clearly becomes a problem as design complexity increases, and as design teams consider incorporating more advanced traditional and formal verification processes within their flow (for instance, cycle-based simulation, two-state simulation, model checking and equivalence checking). Our solution is to introduce a *verifiable subset* of Verilog and a simple RTL coding style. The coding policies we present have enabled us to effectively incorporate these new verification technologies into our design flow. To provide a framework for discussion, we place emphasis on describing verification processes throughout the text--as opposed to an in-depth discussion of the Verilog language. Specifically, we are interested in how an engineer's decision to code their RTL impacts a verification tool's performance and the quality of the overall verification process. Thus, we have deliberately linked the RT-level verification process to the language and have chosen not to discuss the details of the Verilog language reference manual.

In writing and public speaking training, students are always told to know their reader and audience, and adjust their presentation accordingly. In verification, the *audience* for a design description is the verification processes and tools. This book presents the verification process and tools in the first chapters, then presents RTL Verilog in later chapters.

This book tells how you can write Verilog to describe chip designs at the RT-level in a manner that cooperates with verification processes. This cooperation can return an order of magnitude improvement in performance and capacity from tools such as simulation and equivalence checkers. It reduces the labor costs of coverage and formal model checking, by facilitating communication between the design engineer and the verification engineer. It also orients the RTL style to provide more useful results from the overall verification process.

One intended audience for this book is engineers and students who need an introduction to various design verification processes and a supporting functional Verilog RTL coding style. A second intended audience is engineers who have been through introductory training in Verilog, and now want to develop good RTL writing practices for verification. A third audience is Verilog language instructors who are using a general text on Verilog as the course textbook, but want to enrich their lectures with an emphasis on verification. A fourth audience is engineers with substantial Verilog experience who want to improve their Verilog practice to work better with RTL Verilog verification tools. A fifth audience is design consultants searching for proven verification-centric methodologies. A sixth audience is EDA verification tool implementers who want some suggestions about a minimal Verilog verification subset.

The concepts presented in this book are drawn from the authors' experience with large-scale system design projects. The scale of these design projects ranged to more than 200 million gate-equivalents, and we are happy to report that the products were commercially successful. To support the design processes in these projects, we evaluated and integrated verification tools from the marketplace. Before there were commercially available tools, we developed the tools ourselves. The tools include equivalence checkers, cycle-based simulators, linters, implicit interconnection, macro preprocessors, and RTL scan simulation support.

This book is based the reality that comes from actual large-scale product design process and tool experience. We acknowledge that it may not have the pedagogical refinement of other books derived from lecture notes. We go to press with a belief that the timeliness of the principles in this book will prove valuable to RTL design engineers.

Readers can find more information about this book and e-mail the authors from the URL *http://www.verifiableRTL.com*.

Acknowledgments

The authors wish to thank the following people who participated in discussions, made suggestions and other contributions to our *Principles of Verifiable RTL Design* project:

Greg Brinson, Bill Bryg, Christian Cabal, Dr. Albert Camilleri, Dino Caporossi, Michael Chang, Dr. K.C. Chen, Dr Kwang-Ting (Tim) Cheng, Carina Chiang, Dr. James R. Duley, Jeanne Foster, Bryan Hornung, Michael Howard, Tony Jones, James Kim, Ruth McGuffey, Dr. Gerard Memmi, Dr. Ratan Nalumasu, Bob Pflederer, Dr. Carl Pixley, Dr. Shyam Pullela, Rob Porter, David Price, Hanson Quan, Jeff Quigley, Mark Shaw, Dr. Eugene Shragowitz, Dr. Vigyan Singhal, Bob Sussman, Paul Vogel, Ray Voith, Chris Yih, Nathan Zelle, and numerous design engineers from the Hewlett-Packard Computer Technology Lab.

1

Introduction

Myriad books, manuals, and articles have addressed the issue of RTL Verilog style with an emphasis on synthesis-oriented policies. They explain how to write Verilog to wrest optimal gates from the synthesis process. Still other material presents the entire spectrum of Verilog constructs from the architectural specification to switch-level strengths. Yet, these works leave it to the readers to find their way to good practices for verification. Unfortunately, few guidelines govern the coding of RTL Verilog to achieve an optimum flow through the various functional and logical verification processes. This vacuum clearly becomes a problem as design complexity increases, and as design teams consider incorporating more advanced traditional and formal verification processes within their flow (for instance, cycle-based simulation, two-state simulation, model checking and equivalence checking). Our solution is to introduce a *verifiable subset* of Verilog and a simple RTL coding style. The coding policies we present have enabled us to effectively incorporate these new verification technologies into our design flow. To provide a framework for discussion, we place emphasis on describing verification processes throughout the text--as opposed to an in-depth discussion of the Verilog language. Specifically, we are interested in how an engineer's decision to code their RTL impacts a verification tool's performance and the quality of the overall verification process. Thus, we have deliberately linked the RT-level verification process to the language and have chosen not to discuss the details of the Verilog language reference manual (e.g., [IEEE 1364 1995]).

The likelihood and cost of design errors generally rises with complexity. Design errors detected after the product is delivered in large numbers to the marketplace can result in prohibitive cost. In one famous example, an arithmetic design error is said to have cost the company a half-billion dollars, according to Hoare [1998]. He also notes that design errors can have catastrophic effects even in smaller volume products. Examples are launches and improper function of billion dollar satellites, and life-critical system control units in hospitals.

Design errors found prior to customer release are costly as well, leading to long cycles of regression test, inability to run self-test on first silicon, re-spins of silicon, and ultimately, delayed time-to-market. Fortunately, design errors detected early in the design cycle can be corrected far more easily and at a smaller cost than those detected later.

These experiences have forced project managers and design engineers to focus more of their attention on all aspects of verification. Therefore, the performance of design and analysis tools is of paramount concern within any verifiable RTL methodology.

1.1 Register Transfer Level

1.1.1 What is It?

Although the fundamental idea of *Register Transfer Level* (RTL) is universally understood for the most part, there are differences in exact interpretation between different research and development groups. Since RTL is fundamental to this book, starting with the title, it is important that we bring the reader into alignment with the authors' understanding of RTL.

Let us start by examining the three words.

Register. Registers are storage elements (latches, flip-flop, and memory words) that accept input logic states and hold them as directed by timing signals. The timing signals may be clock edges, clock phases or reset. When registers hold multi-bit states, RTL languages declare and reference these state elements either as a vector notation (i.e. a set), or as a bit-by-bit notation.

Transfer. Transfer refers to the input-to-register, register-to-register, and register-to-output equations and transformations. As with the register declarations and references, the equations and transformations operate on vectors as well as bits.

Level. This refers to the level of abstraction. Just as designers of logic systems are more productive by moving from circuit-level design using volts and amperes to Boolean-level zero and one, they become far more productive by designing at the register transfer level operating on the vector-values of sets of bits.

The level in RTL allows for a range of abstraction for equation and transformation notations. For verifiability, we suggest that designer's use the RTL constructs related to the higher-levels of abstraction. For example, where designers can describe state-machines and multiplexers using case, if-else or Boolean, we favor the case and if-else over the Boolean.

Unlike an architectural level of abstraction, the RT-level provides a cycle-by-cycle state-by-state exact correspondence with the gate-level design. The state mapping may be one-to-many or many-to-one, but the correspondence is exact.

Unlike a gate-level of abstraction, for improved verifiability, the RT-level design must not specify timing or other physical characteristics. For example, while the efficient notation of RTL languages make "add" and "and" look the same, in gates a simple n-bit "add" (that would complete its function in one-cycle) will take more time (and area) than a n-bit "and." The focus of RTL abstraction is on the cycle-by-cycle state-by-state correctness.

1.1.2 Verifiable RTL

We define *verifiable RTL* as a combination of coding style and methodology techniques that, when used properly, will ensure cooperation and support for multiple EDA tools used during the course of verification. This cooperation can return an order of magnitude improvement in performance and capacity from tools such as simulation and equivalence checkers. It reduces the labor costs of coverage and formal model checking, by facilitating communication between the design engineer and the verification engineer. It also orients the RTL style to provide more useful results from the overall verification process.

Central to our Verifiable RTL methodology is the concept that the RTL remain the main or *golden model* throughout the course of design. Hence, our functional verification process (e.g., simulation or model checking) can focus its effort on a faster RTL model as opposed to a slower gate-level model. To promote this methodology, the formal verification process of equivalence checking must be used to completely validate equality on all design transformations.

1.1.3 Applying Design Discipline

As Chappell [1999] observes, "designers who get access to the most degrees of freedom will encounter the most anomalies in tool behavior." These designers experience time-to-market costs and delays as they require significantly more EDA tool support.

On one hand, designers using EDA tools should encourage EDA tool developers to be fully compliant with RTL standards. On the other hand, designers should not assume the role of an EDA tool "standards compliance tester" by using every feature described in the standard. While EDA tools will likely improve in future releases, based on compliance testing by users, the design project's costs and delays resulting from tool anomalies will likely damage the corporate bottom line (as well as the designer's career).

Fundamental to the success of today's design projects is a disciplined approach to RTL design. We refer to this as the *Disciplined User Principle*.

Disciplined User Principle
Designers who limit their degrees of freedom in writing RTL will encounter the least anomalies in tool behavior.

1.2 Assumptions

Synchronous Design. As the difficulties in fanning out signals as clock frequencies increase, verification of interacting independent clock domains is becoming more important. This book, however, addresses verification of synchronous designs. This assumption implicitly pervades all of the chapters. Although self-timed asynchronous logic remains an active area of research, RTL verification tool support for clock-synchronized logic domains is far more developed and in widespread use.

EDA Tools. Without discussing specific EDA vendor tool details, this book provides the reader with a comprehensive understanding of various verification processes from a conceptual and practical level. We have deliberately decided not to reference EDA vendor names within the text for two reasons:

- to be fair to all EDA vendors,

- to not date the text with this week's latest EDA vendor acquisition.

A good reference for the latest EDA tools and vendors can be found on the world wide web at the Design Automation Conference's exhibitors list (see http://www.dac.com) and the *comp.lang.verilog* news group.

Applicability. The concepts presented in this book are drawn from the authors' experience with large-scale system design projects. Our discussion and examples draw from actual product design processes and tool experience. Most of the concepts and recommendations discussed are appropriate to designs of any size and complexity. The decision to apply some of the advance concepts (e.g., an object-oriented hardware design pre-processor methodology) needs to be scaled appropriately to accommodate each design project's size, complexity and time-to-market requirements.

1.3 Organization of This Book

Chapter 2 introduces various components of the *verification process* specifically related to verifiable RTL design; such as design specification, test strategies, coverage analysis, event monitoring and assertion checking. In addition, this chapter introduces four important principles of verifiable RTL design, which include the:

- Fundamental Verification Principle,
- Retain Useful Information Principle,
- Orthogonal Verification Principle, and the
- Functional Observation Principle.

Chapter 3, entitled *RTL Methodology Basics*, addresses the problem of complexity due to competing tool coding requirements by:

- introducing a simplified and tool-efficient Verilog RTL *verifiable subset*,
- introducing an Object-Oriented Hardware Design (OOHD) methodology, and
- detailing a linting methodology.

The linitng methodology is used to enforce project specific coding rules and tool performance checks. The principles introduced in this chapter include the:

- Verifiable Subset Principle,
- Object-Oriented Hardware Design Principle, and the

- Project Linting Principle.

Chapter 4 presents the authors' views of the history of *logic simulation*, followed by a discussion on applying RTL simulation at various stages within the design phase. We then discuss how logic simulators work, and how their operation affects simulation performance. Next, we describe optimizations that RTL simulation compilers apply in their translation from Verilog to an executable simulation model. Finally, we discuss techniques for productive application of simulation for design verification entirely at the RT-level. The principles introduced in this chapter include the:

- Fast Simulation Principle and the

- Visit Minimization Principle.

Chapter 5, titled *Formal Verification*, discusses RTL and the formal verification process. We introduce the notion of a finite state machine and its analysis and applicability to proving machine equivalence and FSM properties. In addition, we discuss coding styles and methodologies that will improve the overall equivalence and model checking process. Finally, we illustrate how event monitors and assertion checkers, described in Chapter 2 for simulation, can be leveraged during the formal verification process. The principles introduced in this chapter include the:

- Cutpoint Identification Principle, the

- Test Expression Observability Principle and the

- Numeric Value Parameterization Principle.

Chapter 6 discusses ideas on *verifiable RTL style*, and the reasoning behind them. The style ideas begin with design content, followed by editing practices. The design content includes asynchronous logic, combinational feedback, and case statements. The section on case statements presents the arguments favoring a fully-specified case statement style to facilitate verification.

We then present naming conventions for the various elements of our verifiable RTL style, again with reasoning in support of the style. An important factor in the naming of modules as well as user tasks and vendor library functions is the support of simulation performance profiling, as well as avoiding clashes in their global name spaces during system integration.

The principles introduced in this chapter include the:

- Indentation Principle, the

- Meta-comment Principle, the

- Asynchronous Principle, the
- Combinational Feedback Principle, the
- Code Inclusion Control Principle and the
- Entry Point Naming Principle.

Chapter 7, entitled *The Bad Stuff*, provides specific examples from projects, designers, and EDA verification tool developers that are an impediment to a productive verification process. Compared with other books and papers on RTL design, the most revolutionary ideas in this chapter include classifying the following as *bad stuff*:

- in-line flip-flop declarations, and the
- RTL X-state.

Other *bad stuff* includes:

- RTL versus gate-level simulation differences,
- RTL styles that hamper simulation performance,
- poor design team discipline,
- poor communication between the EDA vendors and the design project, and
- lack of RTL language element policies.

The principles introduced in this chapter include the:

- Faithful Semantics Principle and the
- Good Vendor Principle.

Chapter 8 presents a *tutorial* on Verilog language elements applicable to the register transfer abstraction levels and their verifiable use. For verifiability, we emphasize strong typing, which is not inherently built into the Verilog language, and fully-specified state machines using case, casex and if-else statements. We discuss debugging statements, constant naming, code inclusion controls and command line options for compilation and simulation in a verification environment.

It is the authors' belief that engineers can be far more successful in completing their design by copying and modifying examples. These examples meet the requirements of an entire design flow methodology, and emphasize verification. Formal meta-language specifications of the Verilog language are de-emphasized. Formal specifications are important to Verilog tool imple-

menters concerned with lexical and semantic precision. To the design or verification engineer, however, Verilog examples speak far more clearly than the
legalisms of precise specification.

Chapter 9 draws together and summarizes the twenty one fundamental
Principles of Verifiable RTL Design, which are discussed throughout the
book. We believe that by applying these principles of verifiable RTL design,
the engineer will succeed in adding or improving the use of cycle-based simulation, two-state simulation, formal equivalence checking and model checking
in the traditional verification flow. Furthermore, a verifiable RTL coding
methodology permits the engineer to achieve greater verification coverage in
minimal time, enhances cooperation and support for multiple EDA tools
within the flow, clarifies RTL design intent, and facilitates emerging verification processes. The design project will accomplish a reduction in development
time-to-market while simultaneously achieving a higher level of verification
confidence in the final product through the adoption of a Verifiable RTL
design methodology.

2

The Verification Process

An historical perspective of the design and productivity gain resulting from the progression of one design notation to the next can be viewed as:

1. Schematics (initially used to represent electronic and fabrication details)
2. Boolean equations (Shannon's revelation 1938)
3. Block diagrams and timing charts to handle growing system complexity
4. Flip-flop input equations to a register equation ([Reed 1952])
5. Register Transfer Languages (e.g., DDL [Duley and Dietmeyer 1968])

As these notations progress from one level of abstraction to the next, communicating design intent as well as providing a mechanism to verifying correct functionality are improved [Dietmeyer and Duley 1975].

In the late 1980s, progress in RTL synthesis provided additional productivity gains in turning the designer's notational intent into gates and silicon. Unfortunately, this synthesis productivity gain has resulted in an increase in verification complexity. What we are promoting is a strong coupling of the RTL notation with the verification process. Our focus is on the verification productivity gains resulting from good coding practices. Therefore, to provide a framework for discussion, this chapter (and chapters throughout the book) places emphasis on describing various verification processes. Specifically, we are interested in how an engineer's decision to code their RTL impacts a verification tool's performance and the quality of the overall verification process.

In this chapter, we provide a broad-based discussion on the prevalent challenges with current verification methodologies. These include:

- designing unnecessarily complex block interfaces due to lack of proper specification,
- wasting process time identifying verification targets,
- maximizing verification coverage while minimizing test regressions,
- determining when the verification process is complete,
- observing lower level bugs during the simulation process.

To address these problems, we introduce a simple and effective coding technique for embedding event monitors and assertion checkers into the RTL. This technique is the main focus of this chapter and provides a mechanism for (a) measuring functional coverage, (b) increasing verification observability, and (c) defining verification targets for block-level simulation and model checking.

2.1 Specification Design Decomposition

During the decade of the 1960s, software productivity was dramatically elevated from a low level assembly language coding environment through the inception of higher level programming languages (e.g., FORTRAN, ALGOL). This facilitated the development of interactive, multi-user and real-time systems; which included such complex systems as airline reservations, process control, navigation guidance, and military command and control. Accompanying these new complex software systems, however, were numerous startling problems rarely experienced in the software realm--such as system dead lock, live lock and forward progress problems. To address these problems, the systematic discipline of Software Engineering emerged, providing a management solution to the ever-increasing system complexity [Pfleeger 1998]. Large software system houses abandoned the ad-hoc *Code* $\Rightarrow$ *Design* $\Rightarrow$ *Specify* development model for a verifiable systematic approach to design (i.e., *Specify* $\Rightarrow$ *Design* $\Rightarrow$ *Code*).

It is interesting to compare the events that led to a systematic approach to software design with the state of hardware design today. During the late 1980s, the productivity in hardware design was dramatically elevated from the process of manually generating schematics, to a process of writing a higher level RTL model. At the same time, synthesis technology emerged as an efficient process used to translate the RTL model into a gate-level implementation. This productivity gain has resulted in an increase in design verification complexity. To address these complexities, engineers are beginning to adopt a more systematic and verifiable approach to hardware design--one whose foundation is based on the research, experiences and practice of Software Engineering. Fundamental to the success of this discipline is complete, unambiguous and verifiable specifications.

[Figure 2-1] illustrates the concept of a specification based top-down refinement process utilizing design decomposition. The top levels of the pyramid illustrate what we specified, while the base of the pyramid illustrates how the specification was implemented. This top-down refinement process enables us decompose a higher level of abstraction into a set of lower level components, which can be designed and verified independently [Sangiovanni-Vincentelli et al. 1996].

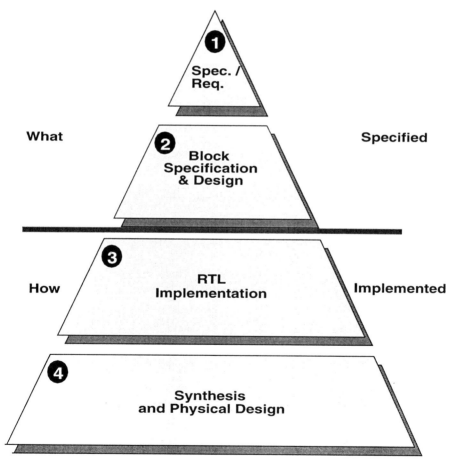

Figure 2-1 Specification Driven Design Decomposition

A common misconception made by many design engineers is reasoning that specification is simply documentation written simultaneously with the RTL code or written as a last step in the design cycle. These designers have a tendency to begin RTL coding prior to fully understanding the block-to-block interface requirements and possible design implementation alternatives. This approach locks the design into specific implementation details too soon while placing unnecessary complexity at the interface between blocks. Thus, this ad-hoc design methodology has a tendency to drive the block interface requirements instead of the other way around. Furthermore, without a clear understanding of the verification target, this ad-hoc design approach limits the concurrent development of tests by the design verification group. Developing an unambiguous specification is fundamental to verifiable design. We summarize the importance of specification by stating the *Fundamental Verification Principle*:

Fundamental Verification Principle

Implementation of RTL code must follow completion of the specification to avoid unnecessarily complex and unverifiable designs.

2.1.1 High-Level Design Requirements

In [Figure 2-1], level one represents the *high-level design requirements*. These requirements typically include functional behavior definition, timing, clock frequency, area, power, performance, as well as software and other hardware interface requirements. Historically, natural languages have been used to describe these high-level design requirements. Keating and Bricaud [1998], however, enumerate limitations with this approach of specifying requirements (e.g., ambiguities, incompleteness, and errors). To resolve these natural language ambiguities, designers typically convert a written specification into more restricted verifiable forms of specification, which include: a *formal specification* (written in a high-level system modeling specification language [Bergé *et al.* 1995]), *executable specification* used as a reference model (usually written in C, C++, or SDL [Ellsberger 1997], or table specification [Eiriksson 1996]). Existing HDLs, such as VHDL or Verilog, are insufficient as specification languages due to:

- the difficulty or inability of expressing environment design assumptions, non-determinism and temporal properties,
- the engineer's tendency to migrate the executable specification toward a lower-level cycle-by-cycle implementation model.

The focus of the executable specification should be on modeling algorithms, not design implementation details.

2.1.2 Block-Level Specification and Design

Level two of [Figure 2-1] represents a refinement of the high-level design requirements into a *block-level specification* through partitioning and decomposition. The block-level specification provides an important link between the high-level design requirements and the RTL implementation. It is through the process of block-level specification and design that we are enabled to explore multiple design alternatives. To reduce implementation and verification complexity, Rowson and Sangiovanni-Vincentelli [1997] assert the premise that a clear separation be maintained between a design's communication (i.e., interfaces) and its behavior at all levels of design refinement (e.g., block and sub-block levels). They point out that an *interface-based design* approach provides many implementation and verification advantages, such as:

1. It partitions a large verification problem into a collection of more manageable pieces, resulting in an increase in verification coverage while improving verification efficiency.

2. It simplifies design and implementation while improving synthesis productivity.

3. It provides a clear interface *contract* between designers of multiple blocks.

4. It facilitates formal verification methodologies.

5. It enables a parallel development of block-level testbenches during RTL implementation.

2.1.3 RTL Implementation

RTL implementation, represented by level three of [Figure 2-1], is a process of refining the block level design specification into a detailed cycle-by-cycle accurate model. The focus of the later chapters in this book is on coding the RTL implementation to facilitate cooperation with all the verification processes within the design flow.

2.1.4 Synthesis and Physical Design

Level four of [Figure 2-1] illustrates the synthesis and physical flow, which represents the process of translating the RTL description into a gate level implementation. To conveniently operate between the RTL description and the physical flow, it is important that the various transformations consider subsequent process requirements within the flow. We refer to this as the *Retain Useful Information Principle*.

Retain Useful Information Principle

A single process within a design flow should never discard information that a different process within the flow must reconstruct at a significant cost.

The *Retain Useful Information Principle* must be considered during all transformation processes involved in the design flow. An example of its application would be embedding the hierarchical RTL signal and wire names in the physical design during flattening. Preserving hierarchical RTL names provides the following advantages for subsequent processes within the flow:

- Permits automatic cross-circuit cutpoint identification using efficient name mapping techniques. Cutpoints, as described in Chapter 5, provides a means for partitioning a large cone of logic into a set of smaller cones. Generally, the equivalence checking process can prove significantly faster a set of smaller cones of logic than one large cone.

- Enables back annotating the place-and-route based scan connections into our RTL (as described in Chapter 3). This yields a 5-10X memory and speed advantage when simulating manufacturing test patterns on the RTL.

- Provides designers with reference points in the physical design when reviewing manual timing tweaks or output from a gate-level simulation. This improves communication between various groups within the design organization.

2.2 Functional Test Strategies

Each hierarchical layer of the specification-driven design process can be validated optimally by defining a clear separation of verification objectives. For example, an executable specification is ideal for validating *algorithmic* requirements and uncovering high-level conceptual problems. Correct ordering of memory accesses (e.g., loads / stores) is a class of properties that are easier to validate at the algorithmic or higher-level verification level than at a lower level of verification.

At the RTL implementation level, ideally a divide-and-conquer and bottom up verification approach should be practiced. This is generally necessary because internal node observability and controllability decrease as the design size increases. For this reason, lower-level RTL implementation properties (e.g., a one-hot state machine will always remain one-hot) are more suitably verified at a block or sub-block level and might otherwise be missed during complete chip or system level verification.

By adopting an interface-based design approach, as suggested in section 2.1, *transaction based verification* can be employed to prove communication correctness when integrating the verified blocks (or sub-blocks). A transaction based verification approach complements a bottom-up verification philosophy. For example, the transaction based verification process can now focus its effort on proving the proper interaction between verified blocks, effectively managing the overall verification complexity. As we move our verification effort up from a block-level to a chip or systems level, a transaction based verification strategy should be employed.

In the following sections, we introduce the most prevalent simulation test strategies used to prove functional correctness at all levels of design. These strategies include *directed, random* and *transaction-based* testing. Typically, directed and random testing strategies will employ either self-checking code or a reference model comparison as the checking mechanism. Transaction-based testing, on the other hand, utilizes rule-based bus monitors. To improve observability, all of these methods can benefit from the use of assertion checkers, which will be discussed in section 2.5.

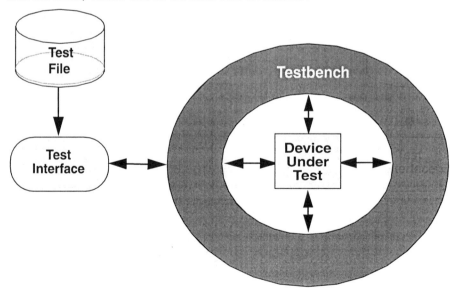

Figure 2-2 Simplified Transaction Analyzer Simulation

2.2.1 Deterministic or Directed Test

Deterministic or *directed tests* are manually written to focus the verification process on particular functional aspects of the design or to increase verification coverage on a specific area of concern (e.g., corner cases). Directed testing, coupled with self-checking code and assertion checkers, is the primary means for validating an executable specification (e.g., a reference model). In general, developing directed tests is labor intensive--even for the simplest test. A more pervasive problem with developing a set of directed tests is anticipating and modeling all the unexpected environment behaviors required for thorough verification.

[Figure 2-2] is an example of a simple testbench environment. This common testing environment allows the direct testing of a block, chip or a combination of blocks and chips. This environment operates on the self-checking test paradigm of (a) providing stimulus for the device under test (DUT), (b) collecting data from the DUT outputs, and (c) comparing the observed data against a set of expected results.

This environment can be constructed to support either cycle-by-cycle or event-by-event testing. The distinction between the two is that for a cycle-by-cycle test environment, the test has an explicit time base reference (e.g., control) for determining when to apply the test stimulus. For an event-by-event test environment, the sequence of operations are applied without an explicit time base reference. This results in a more productive test development environment. The operation of the DUT can be verified to be correct using event-by-event testing; however, the performance characteristics of the DUT are not verified.

2.2.2 Random Test

Directed tests are an effective method for verifying anticipated corner cases. Experience has revealed, however, that the types of defects likely to escape the RTL verification process involve unimaginable and subtle interactions between design blocks or, more likely a complex sequence of multiple simultaneous events. A more appropriate means for finding this complex class of problems (using simulation) is through the use of pseudo-random test generators.

[Figure 2-3] illustrates a simplified random simulation environment. The testbench environment serves the purpose of constraining the device under test inputs to legal combinations. The random stimulus generation block can be of varying complexity (e.g., simple random stimulus, weighted random stimulus, more exotic methods that use feedback from the device under test or testbench environment).

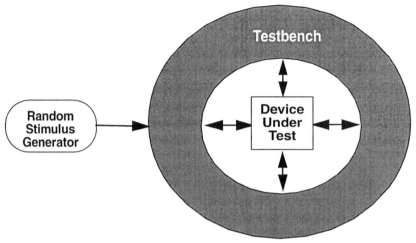

Figure 2-3 Simplified Random Simulation Environment

Alternatively, a pseudo-random test methodology can be implemented utilizing *test case templates*, written by the verification engineer. The test case template specifies fundamental test interaction at a high level of abstraction, allowing a simple template expander tool to unfold the template into a family of test cases--replacing unspecified parameters with random values.

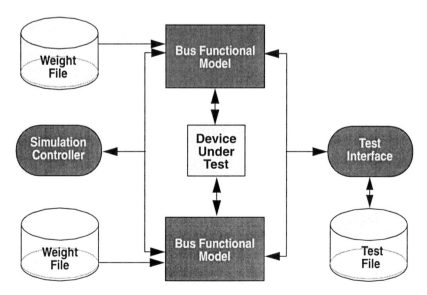

Figure 2-4 Hybrid Random Simulation Environment.

For a complex test scenario, quite often the testbench environment is modeled using a combination of *bus functional models* as shown in [Figure 2-4]. These bus functional models enable the verification engineer to specify the tests as a sequence of transactions at a high level of abstraction, instead of a set of lower level vectors. The bus functional models are used to convert the higher-level transactions into the explicit lower-level signals and operations.

The environment in [Figure 2-4] is referred to as a hybrid random simulation environment, which will support the use of directed and pseudo random generated tests, either individually or collectively. Directed tests are supplied to each bus functional model via the use of a static test input file and the pseudo random tests will be controlled via a bus functional model specific weight file.

The simulation controller manages the initialization of memory, coordinates the test with the bus functional model, and controls the shutdown sequences during error detection. The test execution is coordinated by the simulation controller on a clock-by-clock basis, while the bus functional model interface to the simulation environment is handled on an event-by-event basis. As with our simplified testbench environment example, the simplified random simulation environment is limited to controlling the chip interfaces on specific event boundaries.

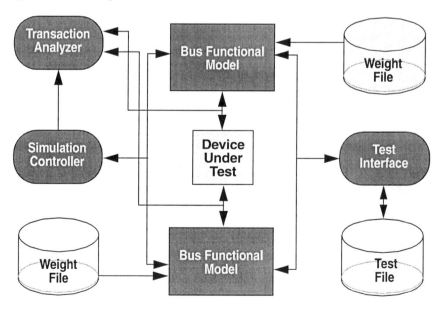

Figure 2-5 Simplified Transaction Analyzer Simulation Environment.

2.2.3 Transaction Analyzer Verification

The simplified transaction analyzer simulation environment, illustrated in [Figure 2-5], is an enhancement to our simplified random simulation environment. It includes the addition of a bus *transaction analyzer*. The transaction analyzer functions as a passive collection of bus monitors, which track and verify data movement throughout the chip or system. For example, an elaborate transaction analyzer could be constructed to keep track of the exact contents of any system memory. In case of a system deadlock, the transaction analyzer is useful for identifying incomplete data transfers--simplifying the debug effort.

2.2.4 Chip Initialization Verification

There are many documented cases, which after many hours of verification, the chip is found *dead-on-arrival* during initial power-up [Taylor et al. 1998][Bening 1999]. One cause for these initialization problems is due to the occurrence of both RTL X-state pessimism and optimism as described in Chapter 7.

Unfortunately, the equivalence checking process will prove that the RTL model and the gate-level model are logically equivalent. Yet, the gate level simulation might uncover an initialization problem that would always be missed due to X-state optimism during RTL simulation. For example, an X value in the RTL will cause the CASE or IF branch to always take the default statement. The engineer, however, might have a functional bug in an alternative branch statement within the RTL. During gate-level simulation, there is no CASE or IF statement X optimism--so the functional bug can be observable if properly stimulated. Hence, the cones of logic are logically equivalent between the RTL and gate-level model, however, the functional bug is not observable at the RT-level. Chapter 7 provides additional details on X-state optimism.

Clearly, strategies for verifying chip initialization need to be considered as a fundamental part of the verification process. Chapter 4 describes a method of using two-state simulation in conjunction with consistent random initialization as a better means for verifying chip initialization at the RT-level. In addition, special consideration needs to be given to running a small set of gate-level simulations earlier in the design cycle to uncover initialization problems.

2.2.5 Synthesizable Testbench

Constraining the testbench HDL to a synthesizable subset appears to be an effective measure. In general, for most chip designs, the RTL description has been successful at preventing race conditions. Testbenches, however, historically have not been quite as successful.

In the *racy* testbench, typically verification engineers tune the races to work with a specific version of a vendor's simulator. This becomes problematic when evaluating competing vendor simulators or new versions of a simulator from the same vendor.

By restricting the testbench HDL to a synthesizable subset and isolating all timing controls into separate modules within the testbench, we are able to prevent race conditions from occurring. Furthermore, the testbench can be moved directly into a cycle-based or emulation environment--significantly improving the overall verification runtime performance. The tutorial in Chapter 8 provides examples of synthesizable testbenches.

2.3 Transformation Test Strategies

It is the authors intent that the RTL remain the *golden model* throughout the course of functional verification. To achieve this goal, its necessary to verify that all transformed models preserve the logical equivalence characteristics of the original RTL model. Formal equivalence checking techniques used to validate design flow transformations will be discussed in detail in Chapter 5.

It is our contention that the verification of functional behavior, logical equivalence and physical characteristics be treated as orthogonal processes, with a clear separation of concerns within the verification flow. We refer to this as the *Orthogonal Verification Principle*. For example, SPICE, static timing verification or gate-level simulation is recommended for verifying physical characteristics (e.g., tri-state, timing, etc.)--while behavioral simulation, RTL simulation and model checking are used to verify functional behavior. Equivalence checking should be used to ensure that the reference RTL model is logically equivalent to a transformed or refined model. In other words, for efficiency and thoroughness, physical characteristic verification is orthogonal to the equivalence checking or functional verification process.

The *Orthogonal Verification Principle* provides the foundation for today's static verification design flows, which enables a verification process to focus on its appropriate concern through abstraction. By applying this principle, we are able to achieve orders of magnitude faster verification, support larger design capacity, and higher verification coverage including exhaustive equivalence and timing analysis.

Orthogonal Verification Principle

Functional behavior, logical equivalence and physical characteristics should be treated as orthogonal verification processes within a design flow.

2.4 Coverage

The objective of design verification is to verify that the design at each level of refinement satisfies the properties of its specification. To guarantee correctness using traditional verification techniques requires enumerating all possible sequences of input and register state, followed by exhaustive simulation and detailed analysis. Although this approach is theoretically possible, it is combinatorially intractable for all but the smallest designs. In practice, formal equivalence checking has been successful in providing 100% coverage when proving equivalence between a RTL reference model and its lower-level transformations (e.g., a gate or transistor netlist). Other formal verification techniques, such as model checking (see Chapter 5), are being used to exhaustively prove correct functional behavior on specific block-level design properties. Unfortunately, this technique do not scale well for high level properties on large RTL models and are not appropriate for proving data path properties within the design. Clearly, a good simulation coverage metric is required, which enables us to measure the degree of confidence in our total verification effort and helps us predict the optimal time for design release (i.e., tape-out). Abts [1999] refers to a set of coverage metrics as the *release criteria*.

Defining exactly what we mean by coverage is a difficult task. Dill and Tasiran [1999] suggest that an objective of verification coverage should be to "maximize the probability of stimulating and detecting bugs, at minimum cost (in time, labor, and computation)." They point out, however, that it is difficult to formally prove that a coverage metric provides a good proxy for bugs, although empirically this seems true. In section 2.4, we explore many different techniques and metrics for measuring verification coverage--and the role coverage plays in contributing to comprehensive validation without redundant effort.

2.4.1 Ad-hoc Metrics

Ad-hoc metrics, such as *bug detection frequency, length of simulation after last bug found*, and *total number of simulation cycles* are possibly the most common metrics used to measure the degree of confidence in the overall verification process. These ad-hoc metrics indicate, after a stable period, that our verification productivity level has diminished. At this point, the verification manager may choose to employ additional verification strategies--or decide to release the design. Malka and Ziv [1998] have extended the use of these metrics by applying statistical analysis on post-release bug discovery data, cost per each bug found, and the cost of a delayed release to estimate the reliability of the design. This technique provides a method of predicting the number of remaining bugs in the design and the verification mean time to next failure (MTTF). Unfortunately, metrics based on bug rates or simulation duration provide no qualitative data on how well our verification process validated the design space, nor does it reveal the percentage of the specified functionality that remains untested. For example, the verification strategy might concentrate on a few aspects of the design's functionality--driving the bug rate down to zero. Using ad-hoc metrics might render a false sense of confidence in our verification effort, although portions of the design's total functionality remain unverified.

2.4.2 Programming Code Metrics

Most commercial coverage tools are based on a set of metrics originally developed for software program testing [Beizer 1990][Horgan *et al.* 1994]. These programming code metrics measure syntactical characteristics of the code due to execution stimuli. Examples are as follows:

- *Line coverage* measures the number of times a statement is visited during the course of execution.
- *Branch coverage* measures the number of times a segment of code diverges into a unique flow.
- *Path coverage* measures the number of times each path (i.e., a unique combination of branches and statements) is exercised due to its execution stimuli.
- *Expression coverage* is a low-level metric characterizing the evaluation of expressions within statements and branch tests.
- *Toggle coverage* is another low-level metric, which provides coverage statistics on individual bits that toggle from 1 to 0, and back. This coverage metric is useful for determining bus or word coverage.

A shortcoming with *programming code metrics* is that they are limited to measuring the *controllability* aspect of our test stimuli applied to the RTL code. Activating an erroneous statement, however, does not mean that the design bug would manifest itself at an observable point during the course of simulation. Techniques have been proposed to measure the *observability* aspect of test stimuli by Devadas *et al.* [1996] and Fallah et al. [1998]. What is particularly interesting are the results presented by Fallah *et al.* [1998], which compares traditional line coverage and their observability coverage using both directed and random simulation. They found instances where the verification test stimuli achieved 100% line coverage, yet only achieved 77% observability coverage. Other instances achieved 90% line coverage, and only achieved 54% observability coverage.

Another drawback with *programming code metrics* is that they provide no qualitative insight into our testing for *functional correctness*. Kantrowitz and Noack [1996] propose a technique for functional coverage analysis that combines correctness checkers with coverage analysis techniques. In section 2.4, we describe a similar technique that combines event monitors, assertion checkers, and coverage techniques into a methodology for validating *functional correctness* and measuring desirable events (i.e., observable points of interest) during simulation.

In spite of these limitations, programming code metrics still provide a valuable, yet crude, indication of what portions of the design have not been exercised. Keating and Bricaud [1999] recommend targeting 100% programming code coverage during block level verification. It is important to recognize, however, that achieving 100% programming code coverage does not translate into 100% observability (detection) of errors or 100% functional coverage. The cost and effort of achieving 100% programming code coverage needs to be weighed against the option of switching our focus to an alternative coverage metric (e.g., measuring functional behavior using event monitors or a user defined coverage metric).

2.4.3 State Machine and Arc Coverage Metrics

State machine and *arc coverage* is another measurement of controllability. These metrics measure the number of visits to a unique state or arc transition as a result of the test stimuli. The value these metrics provide is uncovering unexercised arc transitions, which enables us to tune our verification strategy. Like programming code metrics, however, state machine and arc coverage metrics provide no measurement of observability (e.g., an error resulting from arc transitions might not be detected), nor does it provide a measurement of the state machine's functional correctness (e.g., valid sequences of state transitions).

2.4.4 User Defined Metrics

Grinwald *et al.* [1998] describe a coverage methodology that separates the coverage model definition from the coverage analysis tools. This enables the user to define unique coverage metrics for significant points within their design. They cite examples of user defined coverage metrics targeting the proper handling of interrupts and a branch unit pipe model of coverage. In general, user defined metrics provide an excellent means for focusing and directing the verification effort on areas of specific concern [Fournier *et al.* 1999].

2.4.5 Fault Coverage Metrics

For completeness we will discuss *fault coverage* metrics, which have been developed to measure a design's *testability* characteristics for manufacturing. Unlike programming code coverage, the fault coverage metrics address both controllability and observability aspects of coverage at the gate-level of design. Applying fault coverage techniques to RTL or functional verification, however, is still an area of research [Kang and Szygenda 1992] [Cheng 1993].

These metrics are commonly based on the following steps:

1. Enumerate stuck-at-faults on the input and output pins on all gate (or transistor) level models in the design.

2. Apply a set of vectors to the design model using a fault simulator to propagate the faults.

3. Measure the number of faults that reach an observable output.

2.4.6 Regression Analysis and Test Suite Optimization

By using the various combined coverage metrics described in this chapter, we are able to perform the processes known as *regression analysis* and *test suite optimization*. These combined processes enable us to significantly reduce regression test runtimes while maximizing our overall verification coverage. Using regression analysis and test suite optimization techniques, development labs within Hewlett-Packard have successfully reduced regression vectors by up to 86% while reducing regression simulation runtimes by 91%. Alternatively, Buchnik and Ur [1997] describe a method they have developed for creating small (yet comprehensive) regression suites incrementally on the fly by identifying a set of coverage tasks using regression analysis.

2.5 Event Monitors and Assertion Checkers

In the previous section, we explored various coverage metrics that are used to determine our degree of confidence in the verification process. A deficiency with this standard set of coverage metrics is their inability to quantify functional coverage. In general, verification strategies can be classified as either end-to-end (*black-box*) testing or internal (*white-box*) testing. Properties of the design we wish to validate using white-box testing are easier to check, since it is unnecessary to wait for test results to propagate to an observable system output. In this section, we explore the use of *event monitors* and *assertion checkers* as a white-box testing mechanism for measuring functional correctness as well as detecting erroneous behavior. The use of event monitors and assertion checkers provide the following advantages:

- halts simulation (if desired) on assertion errors to prevent wasted simulation cycles
- simplifies debugging by localizing the problem to a specific area of code
- increases test stimuli observability, which enhances pseudo-random test generation strategies
- provides a mechanism for grading test stimuli functional coverage (e.g., event monitoring coverage)
- enables the use of formal and semi-formal verification techniques (e.g., provides verification targets and defines constraints for formal assertion checkers)
- provides a means for capturing and validating design environment assumptions and constraints

The last point is a notable application of the *Retain Useful Information Principle*. Assertion checkers are a useful mechanism for capturing design assumptions and expected input environmental constraints during the RTL implementation phase. Likewise, event monitors embedded in the RTL provide a mechanism for flagging corner-case events for which the design engineer has verification concerns. The loss of this design knowledge and environmental assumptions can result in both higher verification and maintenance costs.

2.5.1 Events

An *event* can be thought of as a desirable behavior whose occurrence is required during the course of verification for completeness. Examples of events include corner-case detection, such as an error memory read, followed by its proper error handling function. In general, we can classify events as either *static* (a unique combination of signals at some fixed instance of time) or *temporal* (a unique sequence of events or state transitions over a period of time).

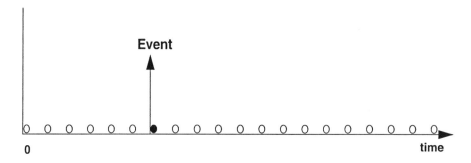

Figure 2-6 Static Event

[Figure 2-6] illustrates the concept of a *static* event, which is a unique combination of signals occurring at some fixed point in time. For example, the static event might be the occurrence of a queue full condition concurrent with a queue *write request*. Creating an event monitor enables us to track this legal corner-case condition, consequently determining whether it has been exercised during the verification process.

[Figure 2-7] illustrates a combined sequence of events exhibiting a unique *temporal* relationship. For example, given that *Event 1* occurs, *Event 2* will eventually occur prior to the occurrence of *Event 3*.

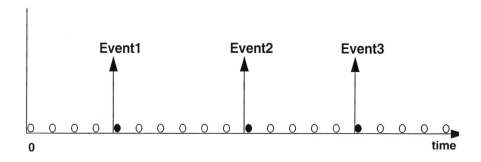

Figure 2-7 Sequence of Events

[Example 2-1] illustrates a simple method of embedding an event directly into the Verilog RTL.

Example 2-1

```
`ifdef EVENT_MONITOR_ON
// Detects when queue is full
// and queue write request occur simultaneously
    always @(c_q_full or c_q_write) begin
      @(negedge ck) begin
       if(c_q_full & c_q_write)
          $display("EVENT%0d:%t:%m", `EV_Q_FULL_WR, $time);
     end
   end
`endif
```

An alternative solution to embedding the event monitor's detection and reporting code directly into the RTL would be to encapsulate the event monitor in a module. This approach provides the following advantages:

1. simplifies the work required by the engineer when specifying events

2. provides clear separation of functional design intent from event monitoring code

3. ensures concurrent detection of events with the execution of the RTL code (as opposed to procedural detection when the event monitor is mixed in with the design's RTL description)

4. permits a seamless optimization of event monitor detection and reporting code throughout design and verification--without disturbing the engineer's text

5. permits a seamless augmentation of new event monitoring features throughout the course of verification

A case in point, the design verification group might decide to augment the [Example 2-1] event monitor reporting mechanism by replacing the simple $display call with a PLI call.[1] This could be used to provide support for a sophisticated event monitor detection scheme that involves a hierarchy of multiple events. By encapsulating the event monitor functionality within a library module, the augmentation of new features can be conducted in a seamless fashion.

---

1. For details on Programming Language Interface (PLI) see [Mittra 1999] and [Sutherland 1999].

[Example 2-2] illustrates a simple method of instantiating an event monitor module directly into the Verilog RTL as opposed to the direct embedded method of [Example 2-1].

Example 2-2

```
// Detects when both queue full
// and queue write request simultaneously occur
event_monitor dv_q_full_write (ck, c_q_full & c_q_write,
                    `EV_Q1_Q2_FULL);
```

The module definition for [Example 2-2] is described in [Example 2-3]. Encapsulating the event detection code within a module permits performance tuning to occur seamlessly, as shown between the clocking scheme differences between [Example 2-1] and [Example 2-3].

Example 2-3

```
`define DELAY_EVENT #2;
:
module event_monitor (ck, test, event_id);
input ck, test;
input [7:0] event_id;
`ifdef EVENT_MONITOR_ON
    always @(posedge ck) begin
        `DELAY_EVENT
        if(test==1'b1)) begin
            $display("EVENT LOG %d:%t:%m", event_id, $time);
        end
    end
endif
endmodule
```

2.5.2 Assertions

Assertion checkers are a mechanism used to enforce design *rules* by trapping undesirable behavior during the verification process (e.g., check for an illegal event or an invalid design assumption). Assertions, like events, can be classified as either *static* or *temporal*. A static assertion can be implemented in a manner similar to that used with an event monitor, with the exception that the event we wish to trap is undesirable behavior.

A temporal assertion can be viewed as an event-triggered window, bounding the assertion. For example, [Figure 2-8] illustrates an assertion check for an *invariant* property. The assertion *P* in this example is checked after *Event 1* occurs, and continues to be checked until *Event 2*. The event-triggers are expressed using any valid Verilog expression. Likewise, the assertion *P* is expressed using a valid Verilog expression.

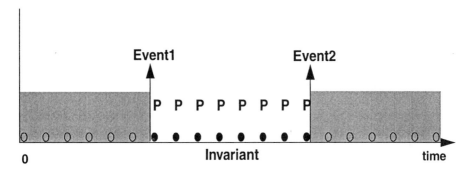

Figure 2-8 Invariant Assertion Window

As another example, [Figure 2-9] illustrates an assertion check for a *liveness* property. The event *P,* in this example, must eventually be valid after the first event-trigger occurs and before the second event-trigger occurs. An assertion error is flagged if event *P* does not occur within the specified event-triggered window.

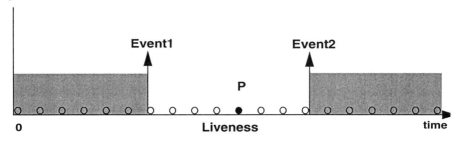

Figure 2-9 Liveness Assertion Window

VHDL provides semantics for specifying static or invariant assertions directly within the RTL as illustrated in [Example 2-4]:

Example 2-4

assert *condition*
report *message*
severity *level*

Verilog, unfortunately, has no equivalent semantics. The techniques we previously presented for monitoring Verilog events, however, are easily extended to validating design assertions. For example, [Figure 2-6] can be expressed using an assertion check as shown in [Example 2-5]:

Example 2-5

```
assert_eventually err_num (ck, ev1_expr, p_expr, ev2_expr,
            `ASSERT_ERR_NUM);
```

In this example, the *Event 1* expression ev1_expr will initiate the monitoring process when it evaluates to true. The assertion checker will then look for a valid p_expr expression prior to the *Event 2* expression ev2_expr evaluating true. If event p_expr does not occur prior to the occurrence of ev2_expr (or the end of the simulation test) then the assertion check fails.

By instantiating a module for the assertion checker, the verification process is able to isolate assertion implementation details from the functional intent of the RTL. This permits the creation of multiple assertion libraries optimized for specific verification processes within the flow. For example, we might create an assertion library optimized for formal verification that contains either vendor specific model checking meta-comments or a library that implements a state machine to trap illegal temporal behavior. Similarly, we could create a simulation-targeted assertion library, which logs events to be post-processed and analyzed for correctness.

An assertion-targeted library should include the following minimal list of assertion checkers:

- **assert_never** -- an event triggered window, bounding a check for an invariant property violation (e.g., an event that should never occur)
- **assert_always** -- an event triggered window, bounding a check for an invariant property violation (e.g., an event that must always occur)
- **assert_eventually** -- an event triggered window, bounding a check for a liveness property violation
- **assert_eventually_always** -- an event triggered window, bounding a check for a liveness property violation (once the property occurs, it must remain valid until the ending event-trigger or the end of simulation occurs)
- **assert_one_hot** -- a check for one hot encoding violations

By setting the *Event 1* expression to 1 and the *Event 2* expression to 0 in our instantiated assertion checkers, the event-triggered window is enabled for all time. For example, an assert_never assertion checker can validate its safety property (p_expr) throughout the entire simulation run when instantiated as follows:

Example 2-6

assert_never safety1 (1, p_expr, 0, clk, `ASSERT_SAFE_1);

[Example 2-7] provides a simplified example for coding the assert_never assertion checker used during simulation.

Example 2-7

```
'define DELAY_ASSERT #2;
module assert_never (ck, event_trig_1, test, event_trig_2,
     serverity_level);
   input ck, event_trig_1, test, event_trig_2;
   input [7:0] serverity_level;
reg test_state;
initial test_state=1'b0;
always @(event_trig_1 or event_trig_2) //assertion window
    if (event_trig_2 II event_trig_1)
      test_state = (~event_trig_2) && (event_trig_1 II test_state);
always @(posedge ck) begin // assertion test
   'DELAY_ASSERT
   if((test_state==1'b1) && (test==1'b1)) begin
     $display("ASSERTION ERROR %d:%t:%m", serverity_level,
         $time);
     $finish;
   end
end
endmodule
```

The tutorial in Chapter 8 provides additional details and examples on assertion checker use. The assert_eventually, assert_eventually_always, assert_always, and assert_one_hot assertion checkers are constructed similarly to [Example 2-7] assert_never checker. Chapter 5 discusses additional details and techniques for extending the use of the assertion checkers into a formal module checking environment. Section 2.5.3 discusses additional considerations and details required to create an effective event monitor and assertion checker methodology.

2.5.3 Event Monitor and Assertion Checker Methodology

Kantrowitz and Noack [1996] describe an effective verification process developed for the DEC Alpha 21264 Microprocessor, which combines assertion checkers, coverage analysis and pseudo-random test generation. Measuring the effectiveness of their various bug detection mechanisms revealed that

34% of all design bugs were identified through assertion checkers (compared with 11% of bugs identified with self-checking directed test). Likewise, [Taylor et. al. 1998] revealed that assertion checkers identified 25% of their total design bugs. Clearly, a verification process that includes event monitors and assertion checkers will quickly identify and isolate errors while improving observability required for functional coverage analysis.

In this section, we discuss a *linting strategy* and *implementation considerations* necessary to form an effective event monitor and assertion checker verification process.

2.5.3.1 Linting Strategy

An effective event monitor and assertion checker linting strategy is required to insure their successful use within the verification process. For example, by comparing different revisions of both the RTL code and event definitions, a simple event linting tool can be created to identify and verify:

- duplicated, new or deleted events
- the absence of events required by a post-processing coverage or verification tool
- signals triggering an event or assertion that have been modified

2.5.3.2 Implementation Considerations

Control. For performance reasons, a monitor enable (or disable) control mechanism must be incorporated within the verification process. [Example 2-1] shows the use of an `ifdef EVENT_MONITOR_ON mechanism to coarsely control the inclusion of monitors at compile time. An alternative mechanism can be implemented, permitting the activation of various classes or sets of monitors at runtime. This allows fine-tuning simulation performance while focusing the verification process on a class of events. For example, a level or classification number could be added as an argument to the event monitor or assertion checker module definition and used for fine tuning control.

Reset and Initialization. Generally, we are only interested in monitoring events or checking assertions after the design has initialized. Determining the status of reset and initialization can be encapsulated within the monitors and checkers to simplify RTL coding. For example:

Example 2-8

if ($dv_initialization_complete()) /// *Check for reset done*
   if(event_expr)   // *Check for event*
     $display("EVENT%0d:%d:%m", `EVENT_NUM, $time);

Text Macro Pre-Processing. A text macro pre-processing flow can greatly reduce the amount of coding required when embedding events and assertions directly into the RTL. This approach permits a seamless augmentation of new features throughout the course of a design project without interfering with the text or functional intent of the designer's RTL. It provides consistency in control and flow--while providing a mechanism for process optimization. From our experience, developing a text macro pre-processor is a minor effort involving a simple *perl* script and small 100 line C program. Isolating assertion checker tool specific details from the design engineer simplifies the task of capturing events and assertions in the RTL while permitting the verification group to tune the methodology at a single point (e.g. the assertion checker targeted library).

[Example 2-9] illustrates a method of coding a text macro directly into the pre-processed RTL file for the liveness check described in [Example 2-5]:

Example 2-9

ASSERT_EVENTUALLY (err_num, ck, ev1_expr, p_expr, ev2_expr,
          `ASSERT_ERR_NUM);

A simple assertion pre-processor can be created that reads in an the text macro source file and generates the final Verilog RTL. The text macros would be expanded in the final Verilog RTL by adding the necessary controls necessary to prevent processing the code during synthesis, along with potentially additional controls for tuning verification performance:

Example 2-10

// *rtl_synthesis off*
`**ifdef** ASSERTION_CHECKER_ON
   assert_eventually err_num (ck, ev1_expr, p_expr, ev2_expr,
         `ASSERT_ERR_NUM);
`**endif**
// *rtl_synthesis on*

As an alternate implementation, the pre-processor could replace the text macro with an inline version of the event monitor or assertion checker as shown in [Example 2-1]. What is important is that a text macro approach permits the verification process to change implementation details seamlessly at a later point in time--while eliminating or minimizing the amount of coding required by the design engineer.

2.5.3.3 Event Monitor Database and Analysis

The development of the Hewlett-Packard PA-RISC 8000 processor provides an alternate example of an event-monitoring methodology. Engineers developing the PA 8000 described events of interest using Boolean equations and timing delays. Rather than embed the events directly into the RTL, however, events were gathered into an simulation event input file and used during the verification process by a simulation event-monitor add-on tool. The list of events the engineers wished to monitor included those that were expected to occur on a regular basis during the verification process, as well as assertion checks that the designer never expects to occur. After simulation, detected events were post-processed and assembled into an *event database*. This database was used to generate *activity reports*, which included statistics such as frequency of events, duration of events, the average, and maximum and minimum distance between two occurrences of events. For details on this methodology, see Mangelsdorf *et al.* [1997].

The event-monitoring methodology used on the Hewlett-Packard PA-RISC 8000 processor development has the advantage of separating the description of events from the simulated code. In other words, the process of event-monitoring is not limited to RTL simulations (e.g., behavior, gate or transistor level event-monitoring could utilize the same process). An advantage with a methodology that embeds events directly into the RTL, however, is that it enables the engineer to capture design assumptions and design knowledge during the development process. Regardless of which event-monitoring methodology is implemented, all methodologies should include a mechanism for creating an event monitor database. Analysis of this database enables us to identify test suite coverage deficiencies and provide valuable feedback on the overall quality of our verification process.

Without looking for specific events and assertions during the course of verification, the designer has no convenient means for measuring functional correctness. The importance of observability and coverage in the design flow is summarized as the Functional Observation Principle:

Functional Observation Principle

A methodology must be established that provides a mechanism for observing and measuring specified function behavior.

2.6 Summary

In this chapter, we examined various components of the verification process specifically related to verifiable RTL design--such as design specification, test strategies, coverage analysis, event monitoring and assertion checking. Furthermore, this chapter introduced four essential principles of verifiable RTL design. We emphasized the importance of specification in the verification process, which we referred to as the *Fundamental Verification Principle*. This principle enables us to reduce design complexity while increasing verifiability. We then introduced the *Retain Useful Information Principle*, which enables us to globally optimize processes within the design flow while capturing environmental assumptions and design knowledge. Next, we discussed the importance of maintaining a clear separation of verification concerns, which provides the foundation for today's static verification design methodology. We refer to this as the *Orthogonal Verification Principle*. Finally, we introduced the *Functional Observation Principle* that is fundamental to the verification process' ability to observe (white-box testing) and measure (coverage) functional behavior.

In addition to the principles presented in this chapter, a simple and effective coding technique for embedding event monitors and assertion checkers directly into the RTL was our main focus. This coding technique provides a mechanism for (a) measuring functional coverage, (b) increasing verification observability, and (c) defining verification targets for block-level simulation and model checking.

3

RTL Methodology Basics

Recent productivity gains in a designer's ability to generate gates have stemmed from the advances and widespread acceptance of synthesis technology into today's design flows. In fact, design productivity has risen tenfold since the late 1980s, to over 140 gates per day in the late 1990s. Unfortunately, design verification engineers are currently able to verify only the RTL equivalent of approximately 100 gates per day. Moreover, designers must now comprehend the relationships, dependencies and interactive complexity associated with a larger set of functional objects, all resulting from the productivity gains of synthesis. Clearly, designers should place proportional emphasis on coding RTL to increase their verification productivity, matching their attention to insuring an optimal synthesis process.

Selecting a single RTL coding style--one that maximizes the performance of simulation, equivalence and model-checking, as well as achieving an optimal flow through synthesis and physical design--is a formidable task. To address the problem of competing tool coding requirements, this chapter introduces three essential RTL coding techniques and methodologies:

1. A simplified and tool efficient verifiable subset of RTL

2. An Object-Oriented Hardware Design (OOHD) methodology, incorporating modern programming language principles

3. A tool-performance-enhancing and design-error linting strategy targeted as the initial check in a complete line of verification tools

This RTL verification-centric methodology creates numerous advantages. It permits a seamless optimization of design processes throughout the duration of the design and enables a seamless augmentation of new processes. In addition, it leaves the designer's text and functional intent undisturbed throughout the course of design. It also offers cooperation and support for multiple EDA tools while achieving higher verification coverage in minimal time. Finally, it clarifies the design intent at the RT level.

3.1 Simple RTL Verifiable Subset

As C. A. R. Hoare [1981] pointed out in his Turing Award Lecture, "there are two ways of constructing a design: One way is to make it so simple that there are *obviously no deficiencies*, and the other way is to make it so complicated that there are *no obvious deficiencies.*"

EDA tool vendors are generally proud of their support of the entire semantic range of the various HDLs. This is as it should be to serve the purpose of selling to a large number of customers. One problem associated with these rich HDLs is that they provide design teams with many alternative methods for expressing the same functionality. Using multiple methods of expression within and across chip designs adds to the overall complexity and therefore cost of verification (in terms of tools, processes and people resources alike).

With the advent of synthesis technology, an RTL *synthesizable subset* has emerged, along with checkers for adherence to a vendor specific synthesis subset. By constraining the rich HDL to a subset of keywords and an explicit coding style, we can use the precise semantic to provide an efficient and clear mapping to a specific hardware implementation.

Other processes within the design flow will also benefit from a constrained subset of the full HDL. For example, the full range of an HDL is supported by event-driven simulation. A faster cycle-based simulation model, however, requires constraining the HDL to a precise subset. The cycle-based simulation HDL subset is generally similar from vendor-to-vendor and similar to the synthesis subset.

The practice of writing HDL that we are advocating in this book is a simple HDL style that serves all verification tools accepting RTL as well as providing a device for communicating clear functional intent between designers. Our justification for advocating an RTL *verifiable subset* are:

1. Vendor tool plug-n-play. It has been our experience that a Verilog *verifiable subset* enables new vendor tools to work "right out of the box", providing an smooth integration path into our design flow. This can be accomplished without the need to re-code the user's Verilog or waiting until a future release of the tool provides adequate language support.

2. Clearer functional intent. Many of the Verilog keywords outside the *verifiable subset* are used to describe lower level switch and Boolean characteristics of the design. Our *verifiable subset* is better suited for describing a higher RT-level of functional intent.

3. Performance. Using keywords outside of the verification subset results in slower RTL functional verification and in many cases precludes the adoption of newer verification technologies, such as cycle-based and two-state simulation, model and equivalence checking.

4. Tool development cost and project schedule. Internal (as well as vendor) tool development cost are significantly higher for each additional keyword supported. A verifiable subset has enabled us to meet project schedules while permitting the successful evaluation and adoption of new verification tools and technologies.

The simplified RTL Verilog subset we are promoting can be measured according to the metrics of keywords, operators, and our "pocket guide."

Keywords. For our *verifiable subset*[1], coding simplicity is emphasized with the use of only 28 out of the 102 Verilog reserved words.

Table 3-1 Verifiable Subset

always	else	initial	parameter
assign	end	inout	posedge
begin	endcase	input	reg
case	endfunction	module	tri
casex	endmodule	negedge	tri0
default	function	or	tri1
defparam	if	output	wire

Typically, minor changes occur between projects, depending on project-specific needs. We believe that it is important, however, to prevent the RTL description from migrating towards lower or gate-level keywords. If it is absolutely necessary to describe physical or implementation level characteristics, then the exceptions should be localized and isolated within their own mod-

---

1. Our Verilog *verifiable subset* is actually a subset of the *synthesizable subset*.

ules. Physical and implementation details should not be embedded directly in with the standard RTL that is used to describe the functional behavior.

The following are 74 Verilog keywords that we recommend be eliminated from a *verifiable subset* and RTL design flow. Most of these keywords specify implementation or lower-levels of abstraction details.

Table 3-2 Verifiable RTL Unsupported Verilog Keywords

and	highz1	rcmos	task
buf	ifnone	real	time
bufif0	integer	realtime	tran
bufif1	join	release	tranif0
casez	large	repeat	tranif1
cmos	macromodule	rnmos	triand
deassign	medium	rpmos	trior
disable	nand	rtran	trireg
edge	nmos	rtranif0	vectored
endprimitive	nor	rtranif1	wait
endspecify	not	scalared	wand
endtable	notif0	small	weak0
endtask	notif1	specify	weak1
event	pmos	specparam	*while*
for	primitive	strong0	wor
force	pull0	strong1	xnor
forever	pull1	supply0	xor
fork	pulldown	supply1	
highz0	pullup	table	

On rare exceptions, a few of the non-RTL keywords missing from our subset can serve in various test bench or system simulation environments. Examples of where these non-RTL keywords appear are:

- ASIC test benches: **force, release, forever**.
- System board level: **supply0, supply1**.
- Gate-level chip model: **primitive**, and many others.

Rare exceptions can be made for looping constructs within library elements or special circumstances by isolating or abstracting the loop into a simple module. To simplify tool support, however, only one of the three Verilog

RTL looping constructs should be selected for a given project.

Recommendation: for Loop Construct

The authors favor the selection of the *for* looping construct over *forever* and *while* if a design project requires an exception to the basic 28 keywords.

Operators. In our verifiable subset, we recommend the use of 30 out of the 35 Verilog expression operators. The remaining five Verilog operators have been excluded from our simplified RTL Verilog coding style.

Table 3-3 Verifiable RTL Unsupported Verilog Operators

operator	example	function
-	-*a*	*unary minus*
*	*a * b*	*multiply*
/	*a / b*	*divide*
= = =	a = = = b	equality (0/1/X/Z)
!= =	a != = b	inequality (0/1/X/Z)

Designers can make an exception with the first three operators when they are isolated from the standard RTL Verilog descriptions.

Chapters 4, 6 and 7 describe the advantages (and techniques) for supporting a two-state simulation methodology. Hence, the last two operators are not required when adopting a two-state methodology.

Pocket guide. Another measure of simplicity for our recommended RTL Verilog coding style is the small 5 page "pocket guide" described in Appendix B, when compared with other much larger full Verilog language pocket guides.

To achieve a verifiable RTL design, we believe it necessary to adopt and enforce a discipline in coding style and RTL subset. We refer to this philosophy as the *Verifiable Subset Principle*.

Verifiable Subset Principle

A design project must select a simple HDL subset, which serves all verification tools within the design flow as well as providing an uncomplicated mechanism for conveying clear functional intent between designers.

3.2 Object-Oriented Hardware Design

Developed in the mid-1960s, the original hardware description languages (HDLs) addressed the deficiencies of building hardware models using first-generation higher-level programming languages.[2] For example, higher-level programming languages during this period lacked support for concurrency and a correspondence or relationship to a specific hardware implementation, at least when modeling at the RT-level. These features are necessary for hardware design analysis and synthesis automation. Today's HDLs have evolved to support the concepts and advancements in modern programming languages as well as the principles that underlie good programming practices (e.g., VHDL [IEEE1076 1993] and Verilog [IEEE1364 1995]). The decision to apply these programming principles when coding in HDLs, however, generally remains up to the engineer.

D.L. Parnus [1972] of Carnegie Mellon University introduced the programming *Principle of Information Hiding*, which instructs designers to hide details pertaining to a single design decision (particularly a design decision that is likely to change) within a program module. This principle enables designers to isolate or localize the design decision details within a module, allowing their interface relationship with other segments of the design to create a level of design abstraction (this is known as the *Abstraction Principle* [Ross *et al.* 1975]).

Engineers are accustomed to creating multiple levels of abstraction within the design's RTL description by partitioning and localizing related functionality into a module description. To optimize each EDA tool's performance, however, we recommend that the *Principle of Information Hiding* be applied (at a minimum) to each distinct grouping of state elements within the design's RTL. This technique allows each register bit (the fundamental building block of an RTL description) to be functionally grouped into a register *object* and detailed at a lower level of abstraction. The process enables the project CAD or design verification group to automatically generate tool-specific libraries, which provide the lower level of abstraction details in an optimized EDA process format. The object's abstraction interface within the RTL description permits referencing a targeted tool specific optimized library for each process point within the design flow. An equivalence checking tool (see Chapter 5) can be used to validate functional consistency between the targeted libraries

---

2. Examples of first-generation high-level programming languages are ALGOL, FORTRAN, COBOL and APL [Iverson 1992]. Examples of the earliest Hardware Description Languages are CDL, DDL, AHPL, and ISP [Chu 1965][Duley and Dietmeyer 1968][Hill and Peterson 1973][Barbacci and Siewiorek 1973].

as shown in [Figure 3-1].

Object-Oriented Hardware Design Principle

Design engineers must code at a higher object level of abstraction--as oppose to a lower implementation or tool specific detailed level--to facilitate verification process optimizations and design reuse.

On the surface, our proposed object-oriented methodology might appear as if we are recommending coding the design at a low gate instance or cell based level. On the contrary, an object-oriented hardware design (OOHD) methodology actually enables coding at a higher level of abstraction. The engineer is now free to design with higher-level objects as oppose to coding lower-level implementation details. In other words, the tool-specific optimization details are now isolated and localized within process specific libraries. Furthermore, the OOHD methodology facilitates design reuse [Keating and Bricaud 1999] by maintaining cell technology independence, which (if required) is controlled within a synthesis-targeted library. The inference of register cells from a library object is still possible during the synthesis process. Alternatively, the explicit referencing of a unique technology cell type within the library object can be controlled during synthesis depending on the user's synthesis optimization requirement

A point the authors would like to make is that developing the object libraries does not require a significant amount of effort, which is based on our experience of developing large systems. The cost of revisiting thousands of lines of RTL code simply to optimize or enable a new verification feature--compared with the cost of developing tools specific libraries--supports our argument for the importance of an object-oriented hardware design approach. Furthermore, our experience has shown that the object pre-processor shown in [Figure 1] is easily implemented with a simple *perl* script and small 100 line C program.[3]

The following sections highlight a few example processes within a typical design flow, and optimization benefits that can be achieved through an OOHD methodology. The specific optimizations described in these examples are not the topics we wish to emphasize. The ability, however, to optimize and tune these and other processes throughout the duration of a project, without inter-

---

3. *Object-oriented programming* focuses on the data to be manipulated rather than on the procedures to do the manipulation. Similarly, *object-oriented hardware design* (OOHD) focuses on a hardware object (such as a 64 bit register, a latch array, or a 32 bit mux) and not the implementation details. The concept of *encapsulation* applies to object-oriented hardware design; however, the object-oriented programming concepts of *inheritance* and *polymorphism* do not apply for our current implementation OOHD using standard Verilog.

fering with the text or functional intent of the original RTL descriptions, is our principal justification for promoting an OOHD methodology..

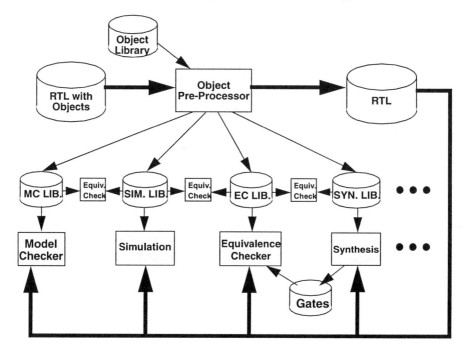

Figure 3-1 OOHD Flow

3.2.1 OOHD and Simulation

The OOHD methodology contributes to a uniform coding style, which enables the functional verification (e.g. simulation or model checking) to remain at the RTL level for the duration of the design process. In addition, the OOHD methodology allows the encapsulation of new verification functionality within a simulation-targeted library later in the design process without disturbing the original RTL description. The next sections describe alternative methods of using the simulation-targeted library to optimize simulation performance.

Simulation Compiler Performance Optimizations. An OOHD methodology favors tuning and optimizing a simulation targeted library for best simulation performance (see [Figure 3-1], *SIM. LIB.*). This is accomplished by providing a uniform coding style within the library to cooperate with the optimizations performed by both event-driven and cycle-based simulation compilers.

For example, the simulation-targeted library can be generated to accommodate an optimization known as *common control consolidation*. By using a standard sequence of control structures for all state elements within the library, the simulator's compiler optimizations is able to maximize its common control structure grouping.

To illustrate *common control consolidation*, consider the following Verilog simple state element assignment with multiplexers, which might be contained in a simulation-targeted library[4]:

Example 3-1

```
module dff (ck,q,d,ck,rst_,scan_sel,scan_in);
    :  // parameterized module
    always @ (posedge ck)

        if (rst_ == 0)
            q <= 0;
        else if (scan_sel )
            q <= scan_in ;
        else
            q <= d;
endmodule
```

Three 16-bit state element objects could be instantiated within the RTL as follows[5]:

Example 3-2

```
dff #(16) dff_a (ck, r_a, c_a, reset_, scan_sel);
dff #(16) dff_b (ck, r_b, c_b, reset_, scan_sel);
dff #(16) dff_c (ck, r_c, c_c, reset_, scan_sel);
```

After an optimizing compiler flattens these three 16-bit grouped state elements from the simulation targeted library; the equivalent Verilog for the assignments would be:

---

4. A synthesis-targeted library might represent the dff object as an actual vendor cell, or higher level RTL code following a synthesis coding practices for register (e.g. if-else statement).
5. Note that in this example the scan input port is left unconnected. Section 3.1.3.2 describes a technique for back-annotating the scan ring connection directly into the RTL.

Example 3-3

```
if (rst_ == 1'b0)
    r_a <= 16'h0;
else if (scan_sel)
    r_a <= \dff_a.scan_in ;
else
    r_a <= c_a;
if (rst_ == 1'b0)
    r_b <= 16'h0;
else if (scan_sel)
    r_b <= \dff_b.scan_in ;
else
    r_b <= c_b;
if (rst_ == 1'b0)
    r_c <= 16'h0;
else if (scan_sel)
    r_c <= \dff_c.scan_in ;
else
    r_a <= c_c;
```

Then, the compiler is able to apply the *common control consolidation* optimization, the final optimized equivalent statements are:

Example 3-4

```
if (rst_ == 1'b0) begin
    r_a <= 16'b0;
    r_b <= 16'b0;
    r_c <= 16'b0;
end
else if (scan_sel) begin
    r_a <= \dff_a.scan_in ;
    r_b <= \dff_b.scan_in ;
    r_c <= \dff_c.scan_in ;
end
else begin
    r_a <= c_a;
    r_b <= c_b;
    r_c <= c_c;
end
```

Note that [Example 3-2] is actually at a higher level of abstraction and easier to code than either [Example 3-3] or [Example 3-4]. If at some future point in time we wish to add a new verification feature or take advantage of a new compiler optimization, encapsulating this new functionality within the dff library module in [Example 3-1] requires less work than visiting each state element in the RTL source when expressed as in [Example 3-4]. Examples of adding new verification functionality include a PLI call to inject errors for high availability testing or a PLI call to support random initialization. Similarly, the OOHD allows us to seamlessly evaluate new EDA or public domain tools that initially require a unique coding restriction much earlier than waiting for the final EDA product Verilog support to stabilize.

Other Verilog simulation compiler optimizations that can be removed from the RTL designer concern and controlled in a simulation-targeted library are:

- *Bus reconstruction.* Occasionally the engineer will group state elements on a bit-sliced portion of a bus. After flattening, the translator optimizer can often combine bit-sliced variable references and operations into full variable references.

- *Flattening.* The translator optimizer flattens (i.e. "inlines") the abstraction boundaries for each module instance within the simulation-targeted library to prevent degradation to simulation performance.

The preceding simulation performance optimizations are examples related to one vendor's simulation recommended HDL style. Simulators from different vendors (as well as successive releases from the same vendor or an in-house simulator) favor unique HDL coding styles for cooperation with their proprietary optimizations. The ability to control a uniform coding style, resulting from an OOHD methodology, allows for continual refinement to accommodate these simulation performance optimizations, without disturbing the designer's HDL.

Two-State Simulator Optimization. In Chapter 4 and 7, we define two-state simulation as the elimination of X from an RTL simulation, and using only 0 and 1 states (and occasionally the Z state). Although tri-state buses have an important place in a system design and simulation, the bulk of the logic and nodes are only 0 and 1, not Z.

During the implementation of a two-state methodology, the verification engineer will generally focus on the improved simulation performance achieved through the elimination of X from our RTL. However, as the verification engineer begins their two-state RTL simulation, they will quickly determine that this method of simulation is a far better verification technique than simulation with an X-state at the RTL-level. For example, Bening [1999], as well as chapter 4, describes the difficulty of identifying start-up initialization or reset problems during RTL-level simulation due to X-state optimism and pessimism.

Two-state RTL techniques that can be encapsulated in a simulation targeted library include:

- Consistent random value initialization across simulators, as well as before and after design changes [Bening 1999].

- Transformation of Z (and X) inputs to random two-state values.

The facility and functionality to support consistent random initialization and Z/X input transformation can be encapsulated within the module descriptions in a simulation-targeted library (e.g., see [Example 3-13]). The support for these two-state verification techniques is an example of adding new process functionality to an existing design flow without disturbing the original RTL when adapting an OOHD methodology.

Key Point. Without an OOHD methodology, the design engineer would be forced on large designs to edit potentially tens-of-thousands of lines of Verilog RTL source code, involving thousands of files, to add optimizations or new verification functionality support.

3.2.2 OOHD and Formal Verification

Chapter 5 describes techniques for applying OOHD to a formal verification design flow. Specifically, equivalence checking is shown to benefit from an OOHD methodology by simplifying the process of latch mapping and master-slave latch folding or compression. Although there are algorithms integrated into commercial tools that automatically identify latch mappings and perform latch folding, applying an OOHD methodology can significantly improve the performance of these tools. Chapter 5 also describes the application of OOHD in a model checking flow, which can be used to solve the multi-phase related clock abstraction problem.

3.2.3 OOHD and Physical Design

To ensure a specific macro-cell implementation during synthesis, particularly related to multiplexers or specific scan-based registers, designers are frequently forced to instantiate vendor specific macro-cells directly into their RTL. By adopting an OOHD methodology, however, the functional verification process is not required to operate on slower gate-level macro-cell models during simulation and model checking. In other words, the synthesis-targeted library (as shown in [Figure 3-1]) allows tuning the lower-level synthesis process for the duration of the project--while a simulation-targeted library maintains a higher level of abstraction view within the RTL.

The following are two areas of physical design, which we have found benefits from our design abstraction and tool-specific library methodology.

3.2.3.1 OOHD Synthesis

There are many details pertaining to developing a company-wide synthesis mythology. One of the challenges is developing a RTL reuse methodology that maintains vendor cell library independence at the RT-level. Our OOHD methodology supports RT-level reuse by:

1. identifying a common set of RTL functional objects,

2. creating a generic Verilog library describing the functional behavior for each object,

3. create, when appropriate, an explicit gate-level version of the generic library to be referenced during synthesis.

An OOHD methodology enables careful control for exact multiplexer (or other) cell selection during synthesis. For many circumstances, correct register inference is still problematic during synthesis. In theory, however, there are good reasons not to force registers to a specific instantiation. For example, it limits synthesis flexibility to select appropriate cell types based on specific optimization time/area compromises.

The reality of day-to-day design is that the designer can usually make the correct choice with only modest extra effort. The benefit is a consistent synthesis result after multiple interations; and a consistent strategy of specifying and implementing clocking, reset and scan across the entire design organization.

Based on our experience using OOHD, groups of register or multiplexer cells can be bundled into an object under one level of hierarchy, which permits referencing the appropriate library for each process within the design flow. For example, in [Figure 3-1], the synthesis targeted library contains the exact instances of desired vendor cells, and is referenced from the RTL during synthesis.

The following is an example of a 20 bit wide 2-to-1 multiplexer macro with an inverted output. It would appear as a pre-process text macro in the RTL as:

Example 3-5

```
MUX2_20 muxes (
    .S (<1 bit port connection>),
    .D0 (<20 bit port connection>),
    .D1 (<20 bit port connection>),
    .X_ (<20 bit port connection>)
);
```

The simulation-targeted library would contain its functional behavior optimized for simulation as follows:

Example 3-6

```
module mux2_20(x_, d0, d1, s);
    output [19:0] x_;
    input s;
    input [19:0] d0;
    input [19:0] d1;
    assign x_ = ~(s ? d1 : d0);
endmodule
```

The synthesis-targeted library would appear as follows:

Example 3-7

```
module mux2_20(x_, d0, d1, s);
    output [19:0] x_;
    input s;
    input [19:0] d0;
    input [19:0] d1;
    mux2_4 m0(
        .s(s), .d1(d1[19:16]), .d0(d0[19:16]),
        .x_(x_[19:16]));
    mux2_4 m1(
        .s(s), .d1(d1[15:12]), .d0(d0[15:12]),
        .x_(x_[15:12]));
    mux2_4 m2(
        .s(s), .d1(d1[11:8]), .d0(d0[11:8]),
        .x_(x_[11:8]));
    mux2_4 m3(
        .s(s), .d1(d1[7:4]), .d0(d0[7:4]),
        .x_(x_[7:4]));
    mux2_4 m4(
        .s(s), .d1(d1[3:0]), .d0(d0[3:0]),
        .x_(x_[3:0]));
endmodule
module mux2_4(s, d0, d1, x_);
    input s;
    input [3:0] d0, d1;
    output [3:0] x_;
    // Vendor Macro Cell
    SLI42M sli42_0(.A1(d1[0]),A2(d1[1]),.A3(d1[2]),.A4(d1[3]),
        .B1(d0[0]),.B2(d0[1]),.B3(d0[2]),.B4(d0[3]),.S(s),
        .X1(x_[0]),.X2(x_[1]),.X3(x_[2]),.X4(x_[3]));
endmodule
```

```
module SLI42M (A1,A2,A3,A4,B1,B2,B3,B4,
  S,X1,X2,X3,X4);
  input S;
  input A1,A2,A3,A4,B1,B2,B3,B4;
  output X1,X2,X3,X4;
  assign X1 = ~(S ? A1 : B1 ) ;
  assign X2 = ~(S ? A2 : B2 ) ;
  assign X3 = ~(S ? A3 : B3 ) ;
  assign X4 = ~(S ? A4 : B4 ) ;
endmodule
```

Section 3.1.4 gives additional details on a possible text macro implementation for an OOHD methodology.

3.2.3.2 OOHD Scan Chain Hookup

Typically, the design's scan ring is connected (or stitched) within the physical design flow. When adapting an OOHD methodology, the wire interconnect for the scan ring can be automatically back annotated into the RTL by including a scan connect stitching file at the top level of the design. This file consists of global references to the scan port connection defined by each of the tool-specific targeted library instantiations. For example, an n-bit parameterized register in the simulation-targeted library might appear as follows:

Example 3-8

```
module dff (q,d,ck,rst,scan_sel,scan_in);
  : // parameterized module
  always @(posedge ck)
  q <= (rst == 1'b1) ? 0 : (scan_sel ? scan_in : d);
  :
endmodule
```

The scan stitch back annotation file, containing global references to the OOHD register instances might appear as follows:

Example 3-9

```
// scan stitch include file -- to be included in the top module of the design
  assign core.in_p3.in_reg.rb0_s.scan_in[3:0] = {
      core.out_m3.out_reg.rb0.q [29],
      core.in_m2.err.r12.q [4],
      core.in_p3.pkt_info.r11.q [7],
      core.out_p3.out_reg.rb0.q [32]};
  assign core.in_p3.in_reg.rb1.scan_in =
      core.in_p3.in_reg.rb0_s.q [3];
  :
```

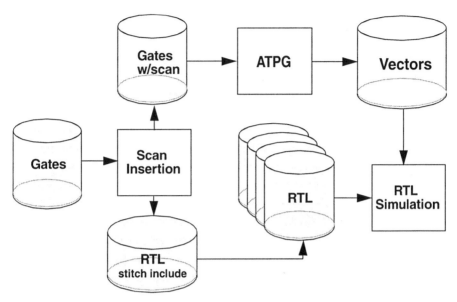

Figure 3-2 Back-annotating scan into RTL.

Notice that the scan_in port, in [Example 3-8], is hierarchically and globally referenced in the scan stitch include file described in [Example 3-9]. Without an OOHD methodology, backannotating the scan connection into the RTL becomes impractical.

[Figure 3-2] illustrates the process of back annotating the scan ring stitch include file back into the RTL In our experience, back-annotating the scan connection into the RTL has enabled us to re-validate the ATPG vectors using RTL simulation (~5 to 10x faster than gate level simulation).[6] Simulating the ATPG vectors on the RTL model enable us to identifying any revision-control or process flow errors, as well as a final functional validation on all libraries used by synthesis, place-and-route and ATPG. These errors might otherwise be missed when equivalence checking at various sequential points within the design flow. In addition, simulating the ATPG vectors on the RTL golden model validates the synthesis tool (and the equivalence checker) by identifying most RTL coding style or interpretation differences between the simulator

---

6. Simulating the ATPG vectors on the RTL is not fault simulation, since there are no gates in our RTL recommended style. From an input, output and register point of view the scan based ATPG vectors should yield the same simulated results between RTL and gates. This is true, provided that all case statements are fully specified and X-state assignments has been eliminated. The advantages for following these coding recommendations will be discussed in Chapter 4 and 6.

and the synthesis tool. Without an OOHD methodology, back-annotating the scan connection into the RTL becomes impractical. With OOHD, however, back-annotation is a trivial effort.

3.2.4 A Text Macro Implementation

There are myriad possible techniques for implementing an OOHD methodology [Bening *et al.* 1997][Barnes and Warren 1999]. One example would be to automatically generate tool-specific libraries during a post process of the RTL by identifying specific instantiations (or objects) requiring a model in the process specific libraries. Conversely, the RTL can be pre-processed and a text macro instantiation used to identify unique objects. Tool-specific libraries are then built along with Verilog RTL expansion of the text-macro as shown in [Figure 3-1]. The pre-process approach has the following advantages over a post-processing technique:

- The unrolling of instance names can be controlled within the synthesis library, which would not be possible using parameterized modules. This is necessary for identifying latch-mapping points between an RTL specification and a gate level implementation.

- Redefining the RTL module port definition is possible for specific process points within our design flow without changing the original RTL pre-process description. For example, adding an enable signal port to support model checking multi-phase related clock abstraction, as described in Chapter 5, is possible without modifying the original RTL pre-process code.

- A boilerplate or template (set of port connection defaults) for commonly used signals within the macro definition (e.g. scan control and reset connections) is easily provided, simplifying the instantiation of objects within the design.

As we previously mentioned, our experience has shown that an object pre-processor is easily implemented with a simple perl script and small 100 line C program. A pre-processing implementation of the OOHD methodology would consist of the following steps:

- Identify all user-specified text macros within the design's RTL by the pre-processor.

- Replace the text macro in the RTL with a unique instantiated module call, creating a new level of abstraction within the design.

- Generate or supplement tool-specific libraries (if needed), which are optimized for specific process points within the design flow.[7]

- Validate the functional equivalence between all tool-specific library created in step 3, using equivalence checking.

[Figure 3-2] illustrates a typical ASIC design flow using the text macro object pre-processor. The *RTL-with-Objects* file is the designer's source RTL, which is written in Verilog with the exception of the various objects, which have been implemented as text macro calls. The *Object Library* contains a macro template description for all targeted libraries within the ASIC flow. For example, [Example 3-10] illustrates the macro template description for a simple non-scan and non-reset based flip-flop, which is targeted for simulation and synthesis, would appear as:

Example 3-10

```
simulation dff_<$1>(q, ck, d);
  output [<$1-1:0>] q;
  input ck;  input [<$1-1:0>] d;
  always @(posedge ck) begin
    q <= d;
  end
  initial $InitialState(q);
synthesis dff_<$1>(q, ck, d);
  output [<$1-1:0>] q;
  input ck;  input [<$1-1:0>] d;
  cffd r<$1-1:0>( .d(d<$1-1:0>),.ck(ck), .q(q<$1-1:0>));
```

The *$1* symbol in the text macro definition is replaced by the macro's formal parameter that specifies the bits size of the state element.

In the body of the *RTL-with-Objects* source file, the designer instantiated their text macro (i.e. object) choices and actual design variables corresponding to the macro's formal parameters.

Example 3-11

```
DFF (2, reg_idle, r_idle, ck, c_idle);
DFF (4, reg_head, r_head, ck, c_head);
```

The pre-processor replaces the text macro references with specific module instances and creates the final Verilog *RTL* file.

Example 3-12

```
dff_2 reg_idle ( .d(c_idle), .ck(ck_a), .q(r_idle));
dff_4 reg_head ( .d(c_head), .ck(ck_a), .q(r_head));
```

---

7. The modules contained within these libraries can be created either uniquely sized or parameterized. The authors favor uniquely sized modules over parameterized modules for reasons described in Chapter 7.

In addition, the pre-processor automatically generates or supplements all tool-specific targeted libraries with unique sized (or parameterized) module definitions. To continue our example, the simulation targeted library, would appear as:[8]

Example 3-13

```
module dff_2(q, ck, d);
  output [1:0] q;
  input ck;  input [1:0] d;
  always @(posedge ck) begin
   q <= d;
  end
  initial $InitialState(q);
endmodule
module dff_4(q, ck, d);
  output [3:0] q;
  input ck;  input [3:0] d;
  always @(posedge ck) begin
   q <= d;
  end
  initial $InitialState(q);
endmodule
```

As previously mentioned, the simulation targeted can be constructed using parameterized modules. However, the Verilog instance name iterater (i.e. parameter iteration on instance names) is not supported in standard Verilog. This is required for efficient name mapping and equivalence checking the final netlist. For example, an optimized synthesis-targeted library for our previous example, which contains the appropriate number of cell instances required for the designer's specified text macro bit width (with uniquely composed instance names), might appear as:

---

8. Note the encapsulation of a user task ($InitialState(q,n)) in [Example 3-13], which provides a mechanism for register consistent random initialization. [Bening 1999b]

Example 3-14

```
module dff_2(q, ck, d);
  output [1:0] q;
  input ck;
  input [1:0] d;
  cffd r0( .d(d[1]), .ck(ck), .q(q[1]) );
  cffd r1( .d(d[0]), .ck(ck), .q(q[0]) );
endmodule
module dff_4(q, ck, d);
  output [3:0] q;
  input ck;
  input [3:0] d;
  cffd r0( .d(d[3]), .ck(ck), .q(q[3]) );
  cffd r1( .d(d[2]), .ck(ck), .q(q[2]) );
  cffd r2( .d(d[1]), .ck(ck), .q(q[1]) );
  cffd r3( .d(d[0]), .ck(ck), .q(q[0]) );
endmodule
```

The equivalence checking process, as described in Chapter 5, can now take advantage of mapping these unique instance names (e.g. r0, r1, etc.).

3.3 Linting

A linting strategy targeted as an initial check in the complete line of verification tools is the most cost-effective method of finding design errors. In addition to identifying syntax errors and project specific coding rule violations, verification-oriented linting makes design and analysis tools--as well as engineering teams--more productive. Hence, a linting methodology must be established early in the design cycle, and used to enforce all verifiable subset and project specific coding style requirements and rules. Ideally, the linting process should be embedded directly into the design flow (in Makefiles, for example), and used to prevent advancing to subsequent processes within the flow upon detection of errors or code style violations. Enforcing a project-specific coding style (*Project Linting Principle*) allows us to achieve a truly verifiable RTL design and is key to our verifiable RTL philosophy.

Project Linting Principle

To ensure productive use of design and analysis tools, as well as improving communication between design engineers, project specific coding rules must be enforced automatically within the design flow.

3.3.1 Linting in a design project.

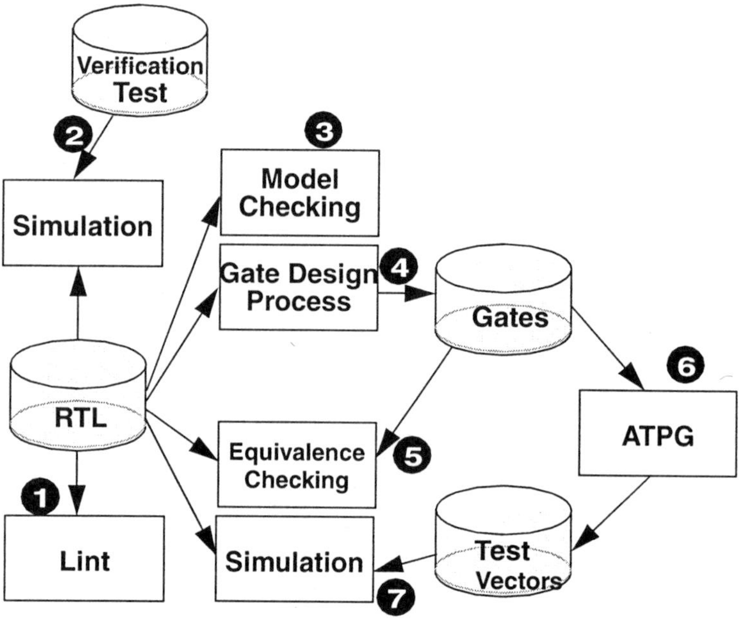

Figure 3-3 RTL-based Verification Flow

[Figure 3-3] illustrates an RTL-based verification flow, which includes linting. The various RTL-based verification flow processes are described as follows:[9]

1. *Linting:* for general-purpose and project-specific rule checks.

2. *Design verification simulation:* including both chip and system simulation, and is either based on slower event-driven simulation models (used for error diagnosis) or a faster cycle-based simulation models (used for error detection). In addition, a faster 2-state simulation model could be used as described in chapter 4.

3. *Model checking:* targeting highly suspect areas where corner cases are not completely known and unlikely to be reached through random simulation test methods.

4. *Implementation Process*: the design process that mixes automatic synthesis plus data path generators, some manual gate design, as well as the physical place and route flow.

---

9. Physical verification is equally as important as functional verification. The authors, however, have chosen to limit their discussion to an RTL-based verification flow.

5. *Equivalence Checking*: used to verify that the refinements or transformations on the design are logically equivalent.

6. *Automatic test pattern generation (ATPG):* used to produce the manufacturing test vectors.

7. *RTL level test vector verification:* Simulation of the manufacturing test vectors against the RTL provides a "belt plus suspenders" double-check of the overall design flow

3.3.2 Lint description

3.3.2.1 Project Oriented

Many of today's commercial linting tools permit customizing *project specific rules* into the general-purpose set of checks. In addition to project specific rules, the following is a list of lint checks aligned with our verifiable RTL design philosophy:

- Strong type checking – e. g. mismatched widths across expression operators are treated as errors.

- Case statement completeness checking - Enforces use of fully specified case statements. See chapter 6 and 7 for additional details and justification of fully specified case statements.

- Project naming conventions – Makes the RTL text better for communication between designers. See chapter 6 for additional details.

- Cycle-based simulation constraints – Includes feedback loop detection, bit-wise check of assignment-before-use in procedural blocks.

- RTL-only subset – Model checking and fast simulation can be compromised by mixing gates with RTL design.

- Clocking restrictions – Enforce project specific clocking conventions.

There are many cases when the design engineer finds it tempting to hand instantiate explicit vendor macro cells directly in the RTL. For example, to secure the timing performance of logic synthesis or to describe a specific implementation for some unique functionality. To prevent loss of the designer's functional intent, and to prevent the RTL code from degenerating into a gate level netlist, the RTL should be written as in [Example 3-1]:

Example 3-15

```
`ifdef IMPLEMENTATION
    <macro cell instance implementation>
`else
    <RTL behavior specification>
`endif
```

For example:

```
`ifdef IMPLEMENTATION
    wire t1, t2, t3;
    XOR3 u1 (t1, in[0], in[1], in[2]);
    XOR3 u2 (t2, in[3], in[4], in[5]);
    XOR3 u3 (t3, in[6], in[7], in[8]);
    XOR3 u4 (perr, t1, t2, t3);
`else
    assign perr = ^in; // calculate parity on 'in'
`endif
```

The preceding example illustrates how a designer might write their Verilog for the gate-level and RTL versions of the same logic within the same module. Chapter 5 describes a methodology for ensuring these two descriptions are logically equivalent.

3.3.2.2 Linting Message Examples

Based on our experience of a good project-specific linting methodology, three classes of lint detection conditions must be established:

- **Errors** - These apply to constructs that would cause incorrect operation, or are not compatible with the various verification processes within the flow (e.g. fast RTL cycle-based simulation or fast boolean equivalence checking). Examples include out-of-sequence statements in procedural blocks, and gate-level constructs (primitives, timing detail) that simulate too slowly.

- **Warnings** - These point to questionable areas of the design that need review before the design can be regarded as complete. Examples include unreferenced bits in a bus and violations of project naming conventions.

- **Advisories** - These apply to simulation performance-related usage. There are typically thousands of these messages about the Verilog for each chip. Designers can optionally ignore them or correct their code for simulation performance as time permits. Typically, designers experience about a one percent improvement in simulation performance for every 100 advisory message that they clear up.

Designers can disable **Warning** and **Advisory** message types by their identifying number. **Error** messages cannot be disabled and are used to prevent the design from advancing to any subsequent process within the design flow.

Here are a few diagnostic linting check examples that we have found useful and have customized into a project specific linting methodology:

Error messages. Although many commercial linting tools detect some of these conditions as a warning, a verifiable RTL design methodology utilizing cycle-based simulation requires reporting these as errors. (NOTE: in the following examples, %s is a valid Verilog identifier or string).

- *Path missing from statement to output*
- *Combinational loop path detected*
- *Identifier '%s' appears both as an assignment target and in the 'always' sensitivity list.*
- *Identifier '%s' appears on both sides of equation.*
- *Identifier '%s' referenced before being assigned.*

Warning messages. These may preclude cycle-based simulation.

- *%s assigned prior to current assign.*
- *Combinational variable '%s' not assigned prior to use in expression*

Advisory messages. These relate to simulation performance. We measured the efficacy of the recommended style by making simulation runs comparing the performance of both styles. In general, the advisory messages try to steer Verilog usage towards references and assignments to an entire bus instead of bit sliced pieces of the bus.

- *Assigning a literal to target '%s' bit or subrange results in slower compiled simulation code; 'or' ('and') variable with a mask '1' ('0') value(s) in the bit position(s) that you need to set (clear).*

Instead of:

    r_b[15:8] <= 8'h00;

use:

    r_b <= r_b & 24'hff00ff;

- *Assigning target '%s' subrange results in slower simulation code. Use concatenation instead.*

Instead of:

```
c_x[23:16] = r_a;
c_x[15:8] = r_b;
c_x[7:0] = r_c;
```

use:

```
c_x = {r_a, r_b, r_c};
```

- *Simulation code for multiple 'xor' operations on selected bits of the same operand as the unary 'xor' ^ operator. Use a mask and unary ^ instead.*

Instead of:

```
c_ecc_out_1 = cin[29] ^ cin[28] ^
        cin[27] ^ cin[26] ^ cin[25] ^ cin[24] ^ cin[23] ^
        cin[22] ^ cin[21] ^ cin[20] ^ cin[19] ^ cin[18] ^
        cin[17] ^ cin[16] ^ cin[15] ^ cin[14] ^ cin[13] ^
        cin[12] ^ cin[11] ^ cin[7] ^ cin[4] ^ cin[1] ^
        cin[0];
```

use:

```
c_ecc_out_1 = ^ (c_in & 40'h003ffff893);
```

- *Simulation code for shift by constant is not as fast as simulation code for concatenation. Use concatenation instead.*

- Instead of:

```
c_v = c_s << 8; // c_v, c_s are 48 bits wide
```

use

```
c_v = { c_s [39:0], 8'h00 };
```

3.4 Summary

In this chapter, we addressed the problem of complexity due to competing tool coding requirements by: (a) introducing a simplified and tool efficient Verilog RTL *verifiable subset*, (b) introducing an Object-Oriented Hardware Design (OOHD) methodology, (c) detailing a linting methodology used to enforce project specific coding rules and tool performance checks.

By constraining the RTL to a *verifiable subset*, the designer will succeed in augmenting their traditional verification process with cycle-based simulation, 2-state simulation, formal equivalence checking, and model checking. A verifiable RTL coding style allows the engineer to achieves greater verification coverage in minimal time, enhances the cooperation and support for multiple

EDA tools within the flow, clarifies RTL design intent, and facilitates emerging verification processes.

By applying the *principle of information hiding* and developing an OOHD methodology, the designer will succeed in isolating design details within tool-specific libraries. This methodology facilitates simultaneously optimizing the performance of simulation, equivalence checking, model checking and physical design within a project's design flow. Furthermore, an OOHD methodology allows adding new features and tools to the design flow throughout the duration of a project, without interfering with the text or functional intent of the original design.

4
RTL
Logic Simulation

The ultimate goal of simulating a design, prior to manufacturing hardware, is to clear out all design errors. This is desirable, to prevent a silicon respin, and provide assurance that the final hardware operates correctly. With the complexities of ever-larger designs, a more practical goal is simulating enough to achieve self-test on the first version of silicon. The worst case scenario is to have the first spin of the hardware lock up at power-on in a state that precludes self-testing.

The challenge in using simulation to achieve self-test, or fully-operational hardware, is that simulation testing is inherently incomplete [Hoare 1998]. Furthermore, the process of simulation makes the validation effort somewhat more incomplete because it is so slow when compared with the actual hardware.

Major design projects of a million gates or more generally run simulation after simulation for months on compute farms consisting of hundreds of CPUs. These simulations run around the clock, 24 hours a day and seven days a week. The total amount of computer time spent in simulation is typically on the order of a million CPU hours. Because logic simulation on workstation and server CPUs runs millions of times slower than the actual hardware, the total of all the simulated clock-cycles typically amounts to one-to-ten seconds of clock-cycles on the actual hardware.

Teams of design and verification engineers craft the simulation test cases. They write directed tests that target basic functionality, so that the first hardware version of the design can achieve self-test. They also develop random tests that try to find unexpected corner cases. To get to the more extreme corner cases, they direct the random cases by skewing the random choices in a direction that stresses the logic architecture. Examples of random stressful conditions processor-based designs might be high cache miss rates, memory addressing bank conflicts, and high no-op counts between instructions. For communication-oriented designs, high transaction rates and severely unbalanced transaction loads may be bring out design errors.

Even though logic simulation is incomplete, design projects find that it is highly productive at finding the numerous simple design errors that occur in the initial version of every design. Designers refer to these as their stupid errors, low-quality errors, or not even errors at all, but mere bookkeeping details, like the use of an active-low signal in place of an active-high signal.

As the low-quality errors are cleared away, and simulation tests begin to pass, design errors become more and more difficult to find through random testing. As the project progresses, increasingly more complex tests are needed to find remaining design errors. Apriori understanding of how to craft tests that make specific errors show themselves becomes more important. At this point, it is possible that an unanticipated design error requiring a complex test sequence may be missed in simulation testing. These are the high quality design errors (brilliant mistakes, made by the best engineers) that must be addressed by the formal verification model checking techniques described in Chapter 5.

Just as formal verification (e.g., model checking) supplements simulation in locating high quality errors, linting supplements logic simulation for the low quality errors. By always linting before running any simulations, a project can clear out design bookkeeping errors more productively than by simulation.

An underlying value that pervades this chapter is the importance of fast simulation, particularly later in the project, when bugs are few and far between. Fast simulation at this phase in the project has the following beneficial results:

- shortens the time to silicon, by reducing the bug rate earlier.
- provides more productive use of the CPU's in the simulation farm,
- locates a few more of those far-between bugs before silicon.

> ### Fast Simulation Principle
>
> *A design project must tailor its RTL and its design process to achieve the fastest simulation possible.*

In section 4.1, we present the authors' views of the history of logic simulation, followed in section 4.2 with how current design projects apply RTL simulation across the design phases. Section 4.3 discusses how logic simulators work, and how their operation affects simulation performance. Section 4.4 describes optimizations that RTL simulation compilers apply in their translation from Verilog to an executable simulation model. Section 4.5 discusses techniques for productive application of simulation for design verification entirely at the RT-level.

4.1 Simulation History

The following history of simulation emphasizes industry application of simulation technology. In some cases, university studies introduced the simulation concepts five to ten years (or more) before they found widespread use in industry computer design shops. In other cases, industry development teams applied ad-hoc methods to solve their problems.

4.1.1 First Steps

Logic simulation began as an idea in the 1950s [Hughes, 1958] when engineers proposed using a simulation model running on a current-generation computer to verify correct function of a proposed next-generation computer design. Designers made little use of logic simulation for design verification, since computers of the 1950s and well into the 1960s had logic gate interconnections in the form of discrete wires. Designers built breadboard prototypes, and made changes to the wiring interconnections when they found a design error. As the design community turned from integrated circuits with a handful of gates to large scale integrated circuits with hundreds of gates, breadboarding for design verification was largely replaced by logic simulation.

Computer design architects in the 1960s used company-specific RTL-like higher-level design notations (above the gate-level) to specify the major design blocks, the data paths, and controls for their designs before building the breadboard prototype. Designers verified their higher-level specifications by their own manual inspection and review by peers. There was generally no computer-based simulation of the higher-level specification.

The first wide production use of logic simulation was gate-level fault simulation, not design verification. These simulators assumed that the design description was good, and determined whether manufacturing faults or field failures in the logic hardware were detectable with a given set of test sequences. The Seshu [Seshu and Freeman 1962][Seshu1965] sequential analyzer would first automatically enumerate all possible stuck-at-one and stuck-at-zero faults for all inputs and outputs on the gates in a design. It would then simulate the effect of these faults, taking advantage of parallel logic operations on the 48-bit word of the host computer to simulate 48 faults at a time.

The simulator in the sequential analyzer used two states, 1 and 0, and compiled the input logic description into host computer assembly language instructions to perform boolean logic operations. The compilation broke feedback loops to form a combinational logic block from sequential logic, and organized the assembly language boolean instructions representing the combinational logic into logic rank order.

In addition to the primary inputs and outputs of the original model, the broken feedback loops in the simulation formed secondary outputs with corresponding secondary inputs. With a given test vector on the primary inputs, the simulator would make an evaluation pass through the combinational logic, then check whether the states of the secondary outputs matched the state of the secondary inputs. If they did not match, the simulator would pass the states of the secondary outputs to the secondary inputs, and re-evaluate the combinational logic. The simulator repeated the evaluations until the secondary output and input states matched, which would indicate that the network had "relaxed", or until a given limit of 100 or so, which would likely mean that there was an oscillation.

Ulrich [1965] described a more realistic simulation time sequencing method to replace rank-ordering. The method mapped cause-and-effect events during simulation, starting at the primary inputs and propagating logic changes through the logic to the outputs. If the logic contained feedback, events propagated (and sometimes created oscillations) through the feedback just like in the hardware. The mapping of time allowed the simulator to deal realistically with logic and wire propagation delays, so that the output of the simulator could look much like the waveforms of logic signals from the actual hardware on an oscilloscope.

4.1.2 X, Z and Other States

Around 1970, authors reported [Bening 1969][Jephson *et al* 1969][Chappell 1971] application of a third "X" state to their logic simulators to model conditions of uncertainty as to whether a logic value was a 0 or a 1. These uncertainties could arise from the start-up state, the outcome of races or min-max delay ambiguity.

Users of gate-level simulators with an X state quickly discovered that the X state simulations erred on the side of pessimism, as described by Breuer [1972]. In spite of the fact that X-state pessimism might lead a designer to fan out a reset signal more widely than really necessary, designers regarded use of the X for a start-up state as a safer design practice than use of an optimistic 0 or 1 initial state.

Some simulators added separate timing-related state values associated with specific transitions, such as rising "U" and falling "D" [Szygenda 1972], which more precisely mapped knowledge of state transitions than using an X for all transitions.

Wilcox and Rombeck [1976] described an extra "Z" state to specify a disconnected node in their simulator. Use of a Z state became widespread in the later 1970s, and continues to this day.

4.1.3 Function and Timing

By the 1980s, static timing verification tools [McWilliams 1980][Hitchcock 1982][Bening 1982] had largely eliminated the need for detailed timing states and delay values in logic simulators. Static timing verifiers largely ignore the logic state values.

If a designer accepts a degree of pessimism, timing verifiers can run in $O(n)$ time and memory complexity, where n is the number of timing blocks in the design. The pessimism arises from a property of timing verifiers that identifies some timing paths as critical where in fact they can never be fully enabled if logic states are consistent. This is known as the false path problem. Designers used to deal with the false path problem by manual analysis and ruling out the offending path, or by case-by-case analysis. Recent developments in static timing verification tools support automatic false path detection algorithms that target failing critical paths.

Even though there is the false path problem with static timing verifiers, they are far more productive than dynamic simulation for timing verification. Simulation for timing verification requires crafting and running of tests for all paths, not just the false paths. It is a very labor and compute-intensive process, with no possibility of completeness on logic designs of a thousand gates or more.

The basis of timing verifiers was built upon the design method of isolating feedback to registers, with no feedback in combinational logic. This was originally described to support testability [Eichelberger and Williams 1977], but serves to support static timing verification as well.

The fact that static timing verifiers focused on timing allowed logic simulation to focus on logic function verification. Design projects using timing verifiers simplified the timing in their logic function simulation, from min-max timing, to unit delay and zero delay. They also turned to larger functional blocks in their simulation, since there was no longer a need for mapping wire interconnect and gate delays into their simulation models.

4.1.4 Gate to RTL Migration

In the 1970s, industry design projects increasingly turned to a register-transfer level description and simulation of their new designs. When the RTL model passed its simulation tests, designers would manually specify the gate-level version of the design, and simulate the tests that had passed at the RT-level. On some projects, designers abandoned the RT-level model once they had a gate-level model that worked.

By the 1990s, synthesis and boolean equivalence checking tools (see Chapter 5) became widely available. Synthesis contributed to the confidence that the gates matched the RTL for the blocks that it could handle. Boolean equivalence proved that the manually crafted gates and the gates that needed timing tweaks matched the RTL as well. This meant that the RTL models could remain the master, and gate-level design verification simulation testing could be eliminated.

4.1.5 Acceleration and Emulation

RTL accelerators based on special architectures have been unsuccessful in the marketplace. Products specifically targeting accelerated simulation beyond the gate-level included Star Technologies STE-264 and the ZYCAD VIP. The STE-264 faded from the scene just a week after its presentation at the Design Automation Conference [Hefferan *et al.* 1985], while the ZYCAD VIP box disappeared in early 1995 with barely any mark left behind.

Gate-level simulation accelerators based on architectures specific to the gate-level simulation application became commercially available around

1984, and experienced rapid sales growth through the 1980s. In the 1990s, gate-level accelerators continue to be commercially available, but their market growth has slowed down.

Earlier simulation accelerators interpreted a machine language instruction set that was comprised of interconnected logic primitive types. Each logic primitive was an instruction to the accelerator, and the instruction addresses were the interconnections. Some simulation accelerators sequenced the evaluation of primitives by use of event lists, and others have used logic rank ordering. [Blank 1984]

From a single-job throughput point of view, these gate-level simulation accelerators provided a significant advantage over the same gate-level simulation run on a workstation. From a price/performance/ease-of-use point-of-view, the advantage was not as good.

Gate-level logic emulators became available in the early 1990s. Emulators can execute logic models at speeds approaching a megahertz. They are about three orders of magnitude faster than simulation accelerators, but are somewhat higher cost and less capacity.

Logic emulators model logic in reprogrammable logic array chips. Modeling logic in logic arrays provides a high degree of gate-level parallelism in execution of the logic model.

In the later 1990s, a new class of accelerators based on arrays of reconfigurable microprocessors has emerged.

To support RTL design, both accelerator and emulator boxes require compilation of the RTL into gate-like box languages. EDA vendors of these boxes provide software support for compiling RTL directly to their box languages, without having to synthesize the RTL to a semiconductor vendor cell library. Later in a project, accelerators and emulators can base their compilation directly on the logic as synthesized into interconnected cells from a semiconductor vendor's library. Simulation performance on accelerators and emulators is the same for a given design described at the RTL or gate-level.

[Figure 4-1] presents approximate relative performance comparisons between RTL simulation running on general-purpose computers and emulator/accelerator boxes. Notice how emulator and accelerator performance is independent of the number of gates, and stays flat until they run out of capacity. RTL simulation declines linearly with increasing model sizes. In our experience performing cycle-accurate RTL simulations up to 450 million gate-equivalents, we have not run into a capacity barrier. This is running on

computers with gigabyte memories and using the cycle-based techniques
described later in this chapter.

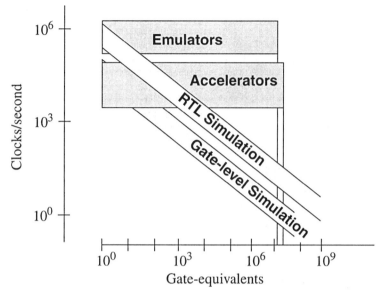

Figure 4-1. RTL Simulation, Emulation, Acceleration Performance
comparisons

Gate-level simulation in software running on general-purpose computers is
one to two orders of magnitude slower than RTL simulation on the same
machines.

4.1.6 Language Standardization

By 1980, there were dozens of different RTL languages [Dewey
1992a][Dewey 1992b]. The RTL languages originating from university-based
researchers had externally accessible papers and documentation about them.
But there was little published about RTL languages originating within industry computer development divisions. In some larger companies, different
computer development divisions separately developed their own RTL languages and simulators. The fact that there were so many different languages
in use in 1980 meant that tool support was language and developer-specific.

There was insufficient formal specification of the semantics of these earlier
RTL languages. The implementation of a logic simulator was the embodiment
of the semantics. Even where some semantic documentation existed outside

the simulator, reports of the simulator not working as documented would result in a change in the documentation.

The syntax was inconsistent within the RTL language in many cases. Within a part hierarchy containing behavioral and structural interconnect modules:

- ports on behavioral modules had one syntax, while ports on a structural interconnect module had another syntax.
- multibit buses were permitted in behavioral modules, and limited in other modules.

VHDL and Verilog development in the 1980s addressed these and many other problems, and became open languages with multi-vendor support.

VHDL started in June, 1981 with a meeting of the Woods Hole Study Group. This select group of 32 HDL language specialists represented about ten different HDLs used by Department of Defense (DOD) contractors and university researchers. There followed a multi-year development process consisting of many meetings and document drafts. By 1987, VHDL was defined, and turned over to the IEEE for standardization. After some changes, IEEE completed its standardization process and released the printed VHDL standard in March, 1988. As part of the DOD contract that funded the VHDL standardization, Intermetrics developed and released the first VHDL simulator.

More simulators and other VHDL-based EDA tools emerged in the market in the following years.

Verilog began with its first customer shipment in 1985 as a proprietary language simulator developed by Gateway Design Automation. Even though its original design target was RTL simulation, the simulation product distinguished itself based on its gate-level simulation performance. Even in design shops that began with an RTL model, there was a need to run regression tests at the gate level.

Gateway merged with Cadence Design Systems, and added tool support for the Cadence language. To sell to customers who favored using an open, standard language with multi-vendor tool support, Cadence put the Verilog language in the public domain in 1990. Cadence Verilog XL™ simulator sales continued their rapid growth, and other vendors developed Verilog tools.

In 1992, Chronologic released their vcs™ "Verilog Compiled Simulator," targeting RTL simulation performance and memory economy, while maintaining close parity at the gate-level with the Cadence simulator.

Our benchmarks showed that, for our designs, Verilog RTL simulation was several times faster than VHDL RTL simulation. Simulation performance has always been a primary value in our design methodology. We regarded the limitations of Verilog compared with VHDL as less significant, and could be overcome. These Verilog limitations include

- weak data type-checking,
- lack of language-based extensibility/configurability, and
- inconsistent evaluation ordering of simultaneous events between different versions of Verilog simulators.

Chapter 3 describes our recommended Verilog RTL methodology to overcome the first two of these limitations.

To design projects that place a higher value on simulation performance than repeatability and use synchronous design methods, the third limitation is not a limitation, but a feature. Leaving evaluation ordering open allows simulator developers opportunities for run time and memory performance optimizations. Synchronous design methods eliminate logic races. With no logic races, any evaluation ordering of simultaneous events results in identical cycle-by-cycle register behavior.

4.2 Project Simulation Phases

Depending upon the amount of design innovation in a new project, the emphasis on the different simulation phases may vary, but generally they consist of:

- *debugging.*
- *regression.*
- *recreating hardware problems.*

To this set of usual simulation phases, we add simulation *performance profiling* between the *debugging* and *regression* phases.

There is overlap between these phases in a large system design project. For example, the chip design simulation models may be largely debugged and in their regression phase, while the system simulation model built from a combination of these chips may be in a debugging phase.

4.2.1 Debugging Phase

As engineers begin their first simulations of their designs, simulations invariably fail to pass their tests due to design bugs. The number of bugs is generally proportional to the size of the simulated design and the amount of innovation applied in the new design.

At this point in the project, an environment that supports productive diagnosis and repair of bugs is of paramount importance. Since the simulations fail their tests in the first 10 to 100 clock-cycles, simulation performance is not as important as the debugging environment. Important components of the simulation debug phase environment include the following:

Access to internal signal values. To track down the causes of design bugs, engineers need to bisect their way anywhere into the design.

Event logging. By including logging about state behaviors that lead up to the point where a design error manifests itself, designers can often more quickly determine the sequence of events that leads to a detected error condition. This can shorten the time that it takes for a designer to diagnose and develop a fix for a design error.

Fast turn-around on design changes. After the engineers diagnose a problem and devise a fix, they need to quickly compile their changed Verilog and re-simulate the test that failed before the change.

4.2.2 Performance Profiling Phase

Before regression testing reaches its peak, it is important to profile the performance of the simulation model and fix any functions that show performance problems. In every simulation that we profile, there are always a handful of functions out of the thousands of functions that stand out as slowing the simulation.

There are always a couple of project-written PLI C/C++ functions in which the algorithms can be changed to greatly improve their performance. Reducing character string operations and converting to block-oriented instead of element-oriented memory management are some of the performance improvement techniques that we have applied based on performance profiling reports.

Most often, our performance problems relate to the simulation output files. Here are some of the output file-related performance problems.

- Writing simulation output files over a network. Use of a disk that is local to the machine on which the simulation is running can give a 5X or more improvement in simulation performance.

- Simultaneously creating multiple simulation output files on a single disk. On multi-CPU machines, use a separate disk for each simulation CPU. This can provide a 1.5X or more improvement in simulation performance.
- Large scale output logging. Where designers ignore most of the logged output, controls on the logged output levels and areas improves simulation performance.

By considering simulation performance important and applying profiling, a project can improve the performance of its simulation model by 10 to 20X over projects that largely ignore performance.

4.2.3 Regression Phase

The regression simulation phase begins when a design passes nearly all of its tests. During this phase, the project uses as much computing horsepower as it can find to run as many different directed, directed random, and random simulations as it can on all the possible configurations of the simulated model.

The rate at which these simulations detect bugs and improve test coverage tapers off in a manner approximating an exponential decay. At some point the bug reporting rate falls off and coverage improvements slow down to a point where the project management team decides that the design is ready to be built into hardware. The timing of this decision is critical to the success of the project. To help them with their decision, the management team applies coverage metrics. See Chapter 2 for more details on coverage metrics.

Later in the regression phase, the verification engineers run thousands of simulations per day and might detect one design error. Because the focus is on detecting errors by running as many simulations per day where nearly all simulations pass their tests, we sacrifice debug support in order to get maximum simulation performance. Here are the ways in which we turn around the priorities for debugging identified in section 4.2.1 when we go into the regression phase:

Limit access to internal signal values. By using and reusing internal registers instead of memory for internal signal values, simulations can run two or three times as fast, and require less memory. Cycle-based [McGeer *et al* 1995] and other optimizations [Ashar and Malik 1995] bypass calculating and storing intermediate combinational logic states.

Reduce event logging. While event logging can help designers diagnose errors detected in simulation, event logging takes simulation time, particularly when it requires a function call. For maximum simulation performance in regression testing, we allow that it may take many cycles before a design error shows itself an abbreviated event log file.

Slower turn-around on design changes. After verification engineers detect a problem in a regression run, they generally have to rerun the same test in order to diagnose the design error and devise a fix. After some preliminary testing in the diagnosis-oriented simulation environment, the designers submit the change to the longer running, more highly optimized compilation for detection-oriented regression simulations.

It should be noted that many of the failing simulations later in the regression phase are not design errors, but rather errors in the test. As verification engineers turn to increasingly complex tests to reveal corner case design errors, it is more and more likely that they have an error in their complex tests than locate a design error in the logic.

Regression RTL simulations continue at a near-peak level even after the project sends out the design data tapes for pre-production prototype chip and board hardware. Design errors detected by simulation can help direct the hardware lab testing, and join what bugs are found in the lab in any respin of the parts.

4.2.4 Recreating Hardware Problems

An important application of RTL simulation is recreating design problems that show up in the hardware lab testing. The first step is crafting the test that duplicates the hardware problem in simulation. The next step is devising the logic change that fixes the problem, then running the test to show that the problem is fixed for that test. The test can then join the regression suite of tests.

Chapter 2 describes in greater detail how projects can focus model checking on the logic area that contained the hardware problem to ensure that the logic change completely fixes the problem.

4.2.5 Performance

With planning, a design project can set up two verification models for simulation of the chips and the system: debugging and regression. The debugging model targets the human user's success, while the regression model targets the CPU performance. [Table 4-1] lists the differences between the two models.

	Debugging Phase	Regression Phase
Verilog compilation	Standard vendor model	Cycle-based optimizations
Internal signal accessibility.	Full accessibility to all registers and intermediate combinational values.	Access to interface buses and selected registers.
Waveform viewing trace file output	Optimized for size and specific waveform viewer.	None.
Diagnostic logging information output	Full diagnostic logging information available	Error detection only
PLI C/C++ code	Debug mode enabled in compile, minimal optimization level.	No debug in compile, maximum optimization levels

Table 4-1. Debugging and regression model differences.

The key element in making the detection and diagnostic models work for a project is the ability to duplicate the behavior of each environment in the other environment.

1. As designs begin passing all of their tests in the slower debugging model and the project moves to the regression, the regression model must pass all of the tests that run correctly on the debugging model.

2. If a regression model simulation run reveals an error running a test, the corresponding debugging model simulation run must duplicate the error behavior running the same test.

Success in making both simulation models match requires that the design project use a modeling style that embodies the same semantics in both the debugging and regression simulation models. Areas to watch for semantic divergence include:

- RTL X-state and two-state. See section 4.5 for details about using random two-state methods to replace the X-state in RTL simulation, and chapter 7 about X-state RTL semantic difficulties.

- Randomly determined test parameters. Since simultaneous event ordering is different between simulation models, randomly generated test parameters must be independent of the order of the calls to the random number generator. This applies to both test direction and start-up state. For test direction, keep the state of the random number generator outside of the random number generator. For more on random start-up state, see section 4.5.1.3.

- Logic races. Where logic contains races, simulation state behavior often diverges as a result of differing simultaneous event ordering between simulations.

Making debugging and regression simulation models run consistently is part of an overall cooperation with all verification tool semantics embodied in the *Faithful Semantics Principle* discussed in chapter 7.

4.3 Operation

To get the best performance from logic simulation, some understanding of the way that simulators work is desirable. Fundamental to logic simulation performance is minimizing the number of *visits*. Visits include signal references and statement/expression evaluations. Visit minimization begins with EDA tool engineers designing logic simulators for efficiency. Design engineers writing Verilog can also contribute to logic simulation visit minimization by the way that they write their Verilog.

Visit Minimization Principle

For best simulation (and any EDA tool) performance, minimize the frequency and granularity of visits.

4.3.1 Sequencing

For the RTL Verilog style described in this book, the sequence of simulation operations consists of alternately evaluating the combinational logic and updating the storage element states. A clock event triggers the storage element state update. Depending on the simulator architecture, the combinational logic evaluation can be sequenced by events during simulation, or rank-ordered during compilation and evaluated in that rank order during simulation.

4.3.1.1 Event-Driven

An event-driven simulator only evaluates logic when input states change value. Evaluations begin with events indicating changes to the inputs. The simulator generates new events for changes to logic states resulting from the evaluations triggered by input state changes. The simulator repeats this process of event propagation until there are no more events in a current time step. The simulator then advances time to the next time step containing events.

Event-driven simulators provide efficiency in two ways.

1. Where a simulation model includes fine-grained delays in combinational logic, the event-evaluate activity per time step is extremely small.

2. Only state changes propagate evaluation visits. Where an event triggers an evaluation of logic that *and*s with a 0 or *or*'s with a 1, the simulator schedules no event triggering evaluation on the output of that logic. When events are sparse, the simulator performs few evaluations.

These advantages of event-driven simulation are now largely obsolete for RTL simulation.

Fine-grained event time management results in simulation performance overhead. Krohn [1981] reported a 52% event management overhead. With static timing verifiers doing the timing verification, logic simulators can focus on logic function verification, and ignore the fine-grained timing.

While some logic designs tend to have low activity in which event propagation gets cut off, other logic designs tend to have high event activity, in which logic blocks may be visited two or more times within a simulated clock cycle. Some sources of high activity and multiple evaluations in event-driven simulation include:

*   Exclusive or's. Exclusive or logic results in high event activity. Exclusive or's propagate every input change to their outputs.

*   Convergent short-paths and long-paths. Events arriving through a short path trigger a visit that causes the logic to switch, while an event arriving through a long path will cause a visit to the same logic, and may cause the logic output to switch again.

Arithmetic logic networks are one example that include both exclusive or events in their short paths, and propagate carry/borrow events along their long paths.

The Verilog in [Example 4-1] illustrates multiple evaluation visits in an event-driven simulation for an evaluation cycle at time 1.

The path from a to e is a short path, while the paths from b and c to e are the long paths. We simulated this model on two simulators from different vendor, and they both reported:

```
        e = 1
        e = 0
```

showing that the simulators visited the e = a ^ d evaluation block twice.

Example 4-1

```
module x;
   reg a,b,c;
   reg d,e;
always @(b or c)
      d = b ^ c;
always @(a or d)          // Extra evaluation in event-driven simulation
   begin
      e = a ^ d;
      if ($time > 0) $display(" e = %b",e);
   end
initial
      begin
      {a,b,c} = 3'b000;
      #1;
      a = 1'b1;
      b = 1'b1;
      #1;
      $finish;
   end
endmodule
```

This example is a near gate-level design. RTL procedural blocks are generally far larger and represent logic function at the conceptual level instead of the boolean function level. For example, RTL addition uses the plus + operator on buses instead of the exclusive-or ^ operator on bits.

RTL procedural blocks may represent all of the combinational logic between registers in many cases. For the cases where combinational logic drives the inputs of larger procedural blocks, multiple visits per cycle can be expensive in terms of performance.

4.3.1.2 Rank-Ordered

Rank-ordering combinational logic prior to simulation greatly reduces the event management overhead during simulation. Compared to gate-level simulation, RTL simulation event overhead is small, while the cost of evaluation visits is usually higher. Eliminating revisit evaluations during a clock cycle is an important benefit that comes from rank-ordered evaluation sequencing.

While it is possible for a logic designer to rank-order manually, a simulation compiler usually performs the rank-ordering. Given the [Example 4-1], we can illustrate rank-ordering by rearranging its two always blocks into one as shown in [Example 4-2].

Example 4-2

```
always @(b or c) begin
    d = b ^ c;
    e = a ^ d;              // Only one evaluation in rank-ordered simulation
    if ($time > 0) $display(" e = %b",e);
end
```

In both standard vendor Verilog simulators and cycle-based simulators, the test sequence shown in [Example 4-2] results in only the display of the final value.

e = 0

[Example 4-3] illustrates a more complex rank-ordering that mixes **assign** statements with an **always** procedural block. Designers need not favor one of these two styles over the other, since simulation compilation optimizations rearrange the original form of the Verilog for performance. Use whatever form most clearly expresses the design from a human point-of-view.

Note that if designers use the larger, multi-statement procedural blocks to specify their logic, they are responsible for placing the statements in rank order. Designers must assign a signal before referencing the signal in an expression. In the [Example 4-3], assignments to c and d must precede the case statement that uses them, and the g assignment that references f must follow the case statement in which f is assigned. Linting checkers are available in the marketplace that flag out-of-sequence statements in procedural blocks as an error.

Example 4-3

a) Original Verilog   b) Rank-ordered Verilog

```
assign c = a;                           always @(a or b or c) begin
assign g = f;                               c = a;
always @( c or d) begin                     d = b;
    case (c)                                case (c)
        2'h0 : f = d;                           2'h0 : f = d;
        2'h1 : f = 1'b0;                        2'h1 : f = 1'b0;
        default : f = 1'b1;                     default : f = 1'b1;
    endcase                                 endcase
end                                         g = f;
assign d = b;                           end
```

4.3.2 Evaluation

With event-driven and rank-ordered logic evaluation sequencing visits, different researchers and EDA vendors have developed simulators using interpreted or compiled-code evaluation methods. In addition, simulator developers have employed evaluation methods specifically targeting RTL statements.

4.3.2.1 Interpreted

Compilers for interpreted simulators generate files of binary address references and instruction codes specifically targeting logic simulation. The simulator consists of an engine that interprets the simulation instruction codes and acts on the codes using selected instructions specific to the host machine.

Interpreted simulation models have the advantages of:

- portability - The file produced by the logic compiler can be independent of a vendor-specific host machine instruction set architecture. Only the simulators interpreting engine needs to target a vendor-specific host machine.
- compactness - the simulator designers optimize the instruction code set for logic simulation.

The big disadvantage of interpreted simulators is that they are slower than compiled-code simulators. Interpreted simulators perform mapping of instruction codes to host machine instructions throughout the entire simulation run.

4.3.2.2 Compiled code

Compilers for compiled code simulators generate host machine instructions that, with the help of a vendor-supplied function library, perform the evaluations required of the simulation model.

Early versions of compiled code simulators generated C or assembly language as an intermediate step in their compilation process, then use the host machine C compiler or assembler to arrive at the object code files representing the design.

The compilers for newer compiled code simulators generate object code files directly, without any intermediate C or assembly language step. These newer compilers use optimization engines and code generators supplied by host-machine hardware vendors, so the code generated is just about as fast as the code that comes from C language compiled with high optimization. The optimization compile time penalty in going directly from Verilog to object code is far smaller (1/10th or less) than the optimization compile time penalty in going from C to object code.

4.3.2.3 RTL Methods

When first learning Verilog, anyone who already knows the C language notices the resemblance of many Verilog constructs as being very much like programming language constructs, particularly within procedural blocks. Since general-purpose simulation host computers optimize their instruction sets and their operation for programming languages, RTL simulation compilers can take advantage of mapping to host computer instructions.

The *transfer* in RTL covers a wide range of abstraction levels that can specify the same clock-by-clock and state-by-state behavior. By specifying designs using language elements at the higher end of the RTL abstraction range, designers can take advantage of host computer instructions that result in higher simulation performance. Here are three ways in which higher levels of RTL abstraction can result in higher simulation performance.

Operators, case and procedural blocks. Instead of boolean expressions in assign statements, favor add/subtract operators (+ and -), case/if-else statements, and procedural blocks to achieve faster simulations, as well as a more clear specification of a design.

Reference buses instead of bits. RTL bus and bit references are another simulation performance factor related to how Verilog statements and expressions map into host computer instructions.Visiting buses instead of bits in the RTL Verilog generally results in faster simulation, as well as simplifying the Verilog.

Throw the X out of your RTL. By throwing out the X state, the RTL to host computer instruction mapping is made even more direct and therefore results in higher simulation performance.

4.4 Optimizations

The optimizations described in this section are only a subset of the RTL optimizations that Verilog compiler writers have put in their software. They continue adding new optimizations in their quest for faster simulation.

An optimizing Verilog compiler may not recognize an opportunity for optimization for some Verilog statements, even though it seems to fit into a class of optimizations. For high usage subblocks, it can be worthwhile in terms of simulation performance to apply some hand-optimization to the Verilog style, and not entirely count on the automatic optimizations doing what you would expect.

4.4.1 Flattening

Flattening eliminates hierarchical boundaries between submodules and the modules that contain the submodules. Flattening can make simulations run faster by eliminating traversal of hierarchical boundaries between submodule port connections.

RTL simulation compilers vary in the degree to which they apply flattening optimizations. Some apply flattening only to smaller, high-usage submodules, and retain the name accessibility to the port connections, even though the port connections no longer exist in the flattened simulation model.

Other RTL simulation compilers aggressively optimize by completely flattening a hierarchy of modules into a single module, and discarding name accessibility to the port connections that they lose in the flattening process. These compilers start their flattening with the top level module in the hierarchy and recursively flatten traversing all of the instances of submodules within the hierarchy.

4.4.2 Rank-Ordering

In section 4.3.1.2, we discussed rank-ordering as an alternative or complementary simulator sequencing option method for event-driven sequencing.

Depending on the simulator, different simulation compilers apply rank-ordering to varying degrees. Some simulation compilers apply rank-ordering to all of the logic in a flattened module, so that the only events are clock events.

4.4.3 Bus Reconstruction

With bus reconstruction optimizations, the RTL compiler combines separate bit and subrange references. These optimizations improve the performance of logic simulation by reducing the number and the granularity of read and write visits.

Note that if subrange assignment statements are far apart in the Verilog, an optimizer may not be able to put the bus back together. Designers can ensure that they get the simulation performance that comes from referencing buses instead of bits by writing their original Verilog in terms of bus references.

4.4.3.1 Concatenation

In many cases, a simulation compiler optimizer can detect and automatically do the subrange assignment-to-concatenation change that turns three bit-wise references into a single full-bus references.

For example, consider optimization of the following source code:

Before: After:

```
c_b [3:0] = c_b0;      c_b = { c_b2, c_b1, c_b0 } ;
c_b [7:4] = c_b1;
c_b [11:8] = c_b2;
```

The <u>After</u> is faster because it only makes one reference to c_b in simulation. (Verilog compilers map signal bits across the bits of simulation host machine words.) Without the concatenation optimization, the simulator has to mortise bits into the assignment target, which requires reads and rewrites of c_b. In Verilog, the work that the simulator has to do looks like:

```
c_b = { c_b [11:4], c_b0 } ;
c_b = { c_b [3:0] , c_b1, c_b [11:8] } ;
c_b = { c_b2, c_b [7:0] } ;
```

4.4.3.2 Expression Simplification

In some cases where the raw flattened Verilog shows that the design uses separate submodule instances to transfer slices from a bus to subranges from another bus. An optimizer can recognize this and simplify the concatenated slices from the same variable into a reference to the entire variable. For example, consider the following code:

Before:	After:
c_c [7:0] = c_a [7:0];	c_c = c_a ;
c_c [15:8] = c_a [15:8];	
c_c [23:16] = c_a [23:16];	

The preceding expression simplification is an RTL-oriented optimization. With the "before optimization" code translated directly to C statement-by-statement, a C optimizer cannot recognize that the goal of all the slicing and shifting is the simple assignment shown in the "after optimization."

4.4.4 OOHD-based Optimization

As first discussed in Chapter 3, use of a design library of standard component written in a uniform coding style results in a large number of common elements in the flattened Verilog. The next two sections discuss optimizations based on expressions and control.

4.4.4.1 Common Sub-expression Consolidation

RTL compiler optimization includes the classic common sub-expression consolidation technique on common statement structures. After flattening the design formed from instantiation of library components, there are thousands of statements of the following form.

```
in_p0.misc.r1.dsel = ( ~ i_scan ) & ( ~ reset ) ;
in_p1.misc.r1.dsel = ( ~ i_scan ) & ( ~ reset ) ;
in_p2.misc.r1.dsel = ( ~ i_scan ) & ( ~ reset ) ;
in_p3.misc.r1.dsel = ( ~ i_scan ) & ( ~ reset ) ;
```

Then, after optimization, there is only one statement to simulate.

```
in_p0.misc.r1.dsel = ( ~ i_scan ) & ( ~ reset ) ;
```

The optimization replaces all of the names that fan out from the common expression with the fanout name in the single remaining statement.

4.4.4.2 Common if-else Control Consolidation

Where designs have multiplexors in front of flip-flops, common if-else control occurs throughout the submodules of an ASIC design. In the flattened

Verilog, these appear as obvious candidates for optimization. Chapter 3 presents an example of simulation optimization based on common control.

4.4.5 Partitioning

Partitioning optimizations improve simulation performance by reducing the amount of logic that the simulator visits in each simulation cycle and fitting evaluated partitions within the simulation host machine cache size. The following sections present RTL partitioning optimizations based on logic branches (case and if), clocks, and chips.

4.4.5.1 Branch Partitioning

[Example 4-4] (a) shows a form of statement sequence that occurs thousands of times in flattened Verilog. Note that **case** output z is used in only one branch of the conditional expression assigned to w. In turn, w is used in only one branch of the **case** that outputs d. When the **case** control variable b is non-zero, all simulation time spent on calculating z and w is wasted.

[Example 4-4] (b) shows the reorganized Verilog after branch partitioning optimizations. Follow the arrows to see how the original statements get moved by the partitioning. These statements only simulate the evaluations resulting in z and w as needed. Branch partitioning improves simulation performance. However, it is sometimes disconcerting to designers when they do not see z and w changing in response to changes in their inputs because they are in the inactive branch of the outer **case (b)** control after optimization.

In addition to bypassing evaluations, the fact that z and w are local variables becomes more clear to downstream optimizations that assign host machine registers instead of memory for local variables.

[Example 4-4] is random control logic from a real chip design, with only the names changed. We find around several thousand lines of these branch partitioning optimization opportunities in control logic out of every 100,000 lines of RTL code. This results in around three to five per cent simulation performance improvement from branch partitioning.

The bigger (2X or more) simulation performance improvement comes in simulation of the scan mode data path. Branch partition in scan mode allows the simulator to bypass evaluation of all the logic in the data path, and only consider the scan path.

Example 4-4

a) Original RTL logic.   **b)** After branch partitioning.

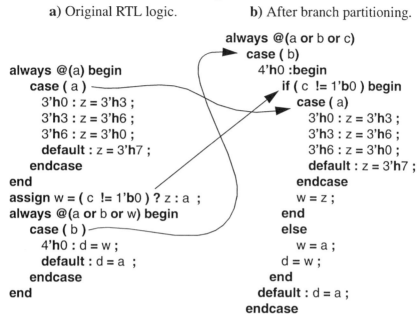

```
always @(a) begin
   case ( a )
      3'h0 : z = 3'h3 ;
      3'h3 : z = 3'h6 ;
      3'h6 : z = 3'h0 ;
      default : z = 3'h7 ;
   endcase
end
assign w = ( c != 1'b0 ) ? z : a ;
always @(a or b or w) begin
   case ( b )
      4'h0 : d = w ;
      default : d = a  ;
   endcase
end
```

```
always @(a or b or c)
   case ( b)
      4'h0 :begin
         if ( c != 1'b0 ) begin
            case ( a)
               3'h0 : z = 3'h3 ;
               3'h3 : z = 3'h6 ;
               3'h6 : z = 3'h0 ;
               default : z = 3'h7 ;
            endcase
            w = z ;
         end
         else
            w = a ;
         d = w ;
      end
      default : d = a ;
   endcase
```

4.4.5.2 Clock Partitioning

Clock partitioning improves simulation performance for designs in which a single master clock arrives at a chip pin, and logic within the chip divides the incoming clock to generate different clock phases. In these designs, the master clock and the clock phases fan out to different flip-flops within the chip.

In designs where some internal clocks fan out only to a few flip-flops, clock partitioning provides a 3X or more simulation performance improvement. The clock partitioning optimization processes chip logic and flip-flops to generate a simulation model that visits and evaluates only the logic and flip-flops affected by each clock.

4.4.5.3 Chip Partitioning

Some RTL simulation compilers flatten a design to the system level, dissolving chip boundaries in multi-chip systems. For larger systems, this method runs into data cache-miss performance difficulties because of random references to the state storage that exceeds the cache size limit.

Application of the *retain Useful Information Principle* [see chapter 2] can tell the RTL compiler which modules are in which chips. The RTL compiler can then generate a system model partitioned by chip instance. This improves the cache-hit percentages and thereby speeds the simulations for multi-chip systems.

At the level of RTL presented in this book, we find that the state-storage memory required per chip for simulation is about 1 byte per gate. Up until now, we have been able to fit the state storage for each chip instance within the cache size available on our host machines. This provides the locality of reference so that simulations can stay in the cache for each evaluation cycle for each chip instance.

If future chip designs require simulation state storage that exceeds data cache limits on future simulation host machines, partitioning to major blocks within each chip will be necessary.

4.5 Random Two-State Simulation Methods

In chapter 7, we describe in detail why RTL simulation with an added X state is not a good idea. In this section, we describe random two-state RTL simulation methods that address start up state and other design problems.

Two-state here refers to eliminating the X, and using only 0, 1 and Z states. Although tri-state buses have an important place in system design and simulation, the bulk of the logic and nodes are only two-state, not tri-state. Most of the following discussion addresses 0/1 two-state simulation, but it does include techniques for treatment of tri-state Z's (or even X's) arriving into a two-state model.

4.5.1 Start Up State

4.5.1.1 Design Method

In their chip logic, designers combine reset signals and chip instance "personalization" input ports to bring their registers to an acceptable start up state.

Some registers do not connect to either reset or input ports. They are designed to be acceptable in any state, or arrive at an acceptable state given a few clock cycles and being fed by the states of the registers that did connect to reset or personalization input ports.

Why not fan out the reset to all the registers?

- Routing area. Reset fanout requires routing area, adding to the cost of physical design, or reducing the total amount of logic that can fit on a given ASIC. Note that for full reset, reset has to go everywhere that the clock goes.

- Timing. Considering the start of reset, reset timing does not immediately seem to be critical. However, the time when the reset signal goes away has to be carefully tuned so that it happens everywhere within the intended clock cycle. Otherwise, some state machines may start "moving" a clock cycle before the others that are still reset. The interaction between them will likely lead to bad outcomes.

- Design verification test. For some free-running counter registers, any start-up state should be acceptable.

4.5.1.2 Zero Initialization

Bening [1999b] described a project that turned to two-state RTL simulation when their ASIC designs started passing their tests using the vendor Verilog RTL simulation model. The project began their simulations with registers in an X state.

All of their chips failed their tests using a cycle-based two-state simulation that started with registers in an zero state. They first suspected that there was a bug in their cycle-based two-state ASIC simulation, because it was newer and not as widely used on different designs as the vendor simulator.

To verify this, they added PLI calls to the chip RTL Verilog that initialized all registers and memory arrays to zero, and simulated them using the vendor simulator. They all failed their tests!

The problems were in the designs and not in the cycle-based simulator. What the designers found was that they had tuned their RTL designs to simulate with registers initialized to X. Their **if** and **case/casex** statements tested their control variables against two-state constants. With registers initialized to X, in the first simulation cycles if-else statements took the **else** branch, **casex** statements took their first branch, and **case** statements took their **default** branch.

With registers initialized to 0, wherever **if-else** and **case/casex** statements compared control variables with zero, the statement took that branch.

In effect, initializing registers to zero amounted to a different test for the ASIC than initializing to X.

Designers found that design problems brought to light by initializing registers to zero were relatively easy to track down, compared with whatever gate-level X-initialization problems they had faced in the past. This was a result of working with "real" 0 and 1 states that occur in the hardware, not the X, which only occurs in simulation.

4.5.1.3 Random Initialization

If RTL zero initialization can find more bugs in RTL simulation than the X state, it is intuitive that random initialization is even better. The same PLI calls that initialize to zero can initialize storage elements to random 0 and 1 states.

The PLI functions can use a time-of-day or user-specified seed for generating random values. When using the time-of-day seed, the PLI functions must report that seed, so a user can re-create any problems detected with the time-of-day seed by specifying that seed in subsequent simulations.

Note that instead of spraying the design with random state by a single PLI function that traverses the design tree, our OOHD methodology [described in Chapter 2] supports each library module taking care of its own initialization in a separate PLI function call.

Spraying the design with random state by traversing the design tree works all right with an event-driven four-state simulator in which registers start at X. However, it does not work in two-state. In two-state event-driven simulation where registers start at 0, many values inserted into combinational variables driven by registers starting at 0 will be inconsistent with the register that drives them. The inconsistency happens wherever a registers starts at 0 and gets set to 0. Unlike starting at X, there will be no event to propagate to make the values of the combinational variables that the register drives consistent.

An important random state initialization feature is consistent random initialization[1]. The consistent result aspect is important for two reasons. Given the same seed, a project needs to duplicate the machine state for a design problem:

---

1. Hewlett-Packard patent pending.

1. between the cycle-based chip simulation model and the vendor-based chip simulation model. A design problem detected by detection-oriented simulation model could be re-created and diagnosed by the other simulation model, given the same seed.

2. after a design change intended to fix the problem. This allows designers to verify that a design problem for a given state had been fixed after the design change. Given the same seed, all register state bits start in the same state as before the design change, except for new register bits that were added in the with the design change.

4.5.1.4 Verification Test

One important fact to note is that by NOT fanning out reset everywhere and NOT scanning a defined state into every register bit, random state initialization becomes a design verification test feature in some cases.

If a project designs their chips in a way that all register states were defined at start up, it might take many clocks of a directed or random test sequence for the interacting state machines to arrive at states that manifests a design problem in their interaction.

By leaving registers open to random states at start up, a project can find some state machine interaction design problems at the beginning of a simulation.

One example of this is a refresh counter. With each run of a simulation test, the timing of the refresh cycles interrupts the test sequence at different times. As the test passes for each run, it provides additional assurance that the interacting state machines in the design can handle the test refresh interruption correctly.

4.5.2 Tri-state Buses

To simulate tri-stated buses in a system design, a project can set up boundaries between the two-state simulation regions and the tri-state signal lines. The tri-state signal lines can use 0, 1, Z and X values, while the remainder of the system signals uses only 0 and 1. The bus drivers consist of a two-state enable signal, and the two-state signal being enabled.

The tri-state bus receivers have to deal with the situation of interfacing the potentially inactive bus signals at a Z state with two-state logic and registers. A correct logic design would not enable the bus at a Z state into any two-state region, but we want to somehow expose the situation where a design problem caused the bus signal receiving logic to be active when the bus signal was at a Z state.

For the same reasons (simulation performance, labor content, complexification, completeness, synthesis) we do not extend our RTL Verilog to watch for the X state in if-else and case statements, we do not extend them for the Z state either.

Using the OOHD techniques described in chapter 2, we add a Z state trapping PLI call right at the tri-state to two-state bus receiver boundary. So, given a tri-state pin on one side of the boundary, and a two-state ipin on the other side, the

$TrapXZ(ipin,pin,qpin);

function passed through 0 and 1 values on pin to ipin, but put random 0 and 1 values from qpin onto ipin for any bit(s) that were Z (or X) on pin.

The intent of the random values is to cause design problems to manifest themselves because of bad data or invalid control signals.

4.5.3 Assertion Checkers

It is possible to envisage assertion checkers as an alternative or complementary method to two-state simulation and inserting random values.

Here are some examples:

- the default in a case statement that checks for all possible two-state values could issue a diagnostic message when the case control variable has X bit values.
- bus Z-states arriving at an active input

Compared with assertions, design problems expose themselves differently when found by random values. The random values at start up and substituted for Z-state signals coming into a two-state register introduce data path parity errors and control state machine sequence malfunctions. Though a little harder to diagnose than an assertion that "talks to" the test engineer, the random values provide more complete coverage with less labor content than assertions.

As with other verification methods that may overlap in terms of what they can detect and diagnose, assertion checkers are acceptable in a two-state methodology. However, it should be noted that we do not allow X-detecting assertions from our RTL style.

4.5.4 Two-State in the Design Process

Bening [1999b] reported that combining:

- $0.5 * 10^6$ RTL two-state simulations randomly initialized with different seeds combined, and
- 500 gate-level simulations starting at X

were sufficient for completely eliminating start-up state design problems in the first silicon. The RTL two-state simulations detected 18 start-up state design bugs, and the gate-level simulation starting at X detected five more.

In our experience on prior chip design projects, X-state RTL simulation masked start-up state problems that made it to silicon. Zero and random two-state initialization would have caught those problems in simulation, according to the designers on those earlier ASIC projects.

Describing the DEC Alpha functional verification methods and experience in [Taylor *et al.* 1998], the authors mentioned a bug that got through to silicon due to insufficient randomization of the RTL simulation model initial state. The authors did not quantify *insufficient*.

Bening [1999b] reported that unlike other HDL dialect and policy recommendations where there was sometimes been lingering disagreement, acceptance of the two-state simulation HDL style among the design team was one hundred percent. All it took to convince the designers was the experience of simulating a design that passed tests starting at X, and failed tests starting at zero.

Fundamental Rule: Simulate RTL two-state

RTL chips and systems must be designed to simulate correctly with a two-state simulation model.

4.6 Summary

Even though RTL simulation is slow compared with the actual hardware that it models, it is highly productive in finding the many simple design errors that invariably exist in new designs. Fast RTL simulation can help a project's schedule, find more design errors and use simulation host machines more productively.

The history of simulation has carried us to a point where we now have standard design languages with broad tool support. Static timing verifiers and

RTL-to-gate boolean equivalence tools now allow design projects to use RTL simulation for all of their design verification simulation work.

Design projects can be most productive by employing an RTL style that targets both efficient debugging simulations as well as fast regression simulations. Achieving this kind of productivity early and later in the project requires consistent RTL semantics between the regression and debugging simulations.

Some understanding of how simulators work can point a designer writing Verilog towards RTL language elements that simulate faster, as well as clarifying the designer's intent. Examples include reducing the number and granularity of evaluation visits to statements and use of statements that describe the design at higher RT levels.

Similarly, understanding of RTL simulation logic compiler optimizations suggests RTL styles by which the logic designers can express their designs already pre-optimized.

Attention to RTL language elements and styles that simulate fastest is particularly important in high-usage library elements. Where there is doubt about which of several alternative styles that simulate faster, library component developers should compare the simulation performance using a setup like that shown in appendix A.

Two-state simulation provides a faster regression simulation model than simulation using an X state. Random start up state initialization supports RTL Two-state verification that is superior to X state for common case-default and if-else constructs. Use of two-state and random simulation methods in a regression environment requires that debugging simulations can replicate problems detected in regression simulations. The random simulation replication methods include both random test direction and random start up state.

5

RTL
Formal Verification

In Chapter 2, we introduced the **Orthogonal Verification Principle**, which states that functional behavior, logical equivalence and physical characteristics should be treated as orthogonal verification processes within a design flow. With increasingly complex designs, we need to spend more time verifying functionality and automate the process of verifying design transformations (e.g., synthesis and physical design). In this chapter, our goal is to introduce a technique that enables us to separate the verification of circuit equality vs. circuit functionality. Furthermore, to address the verification coverage concerns of traditional simulation, we introduce the technique of state-space exploration focussed on the RTL (e.g., model checking).

Equality. Historically, proof of equality has been a challenge within the design flow. For example, as the design process transforms the RTL model into a physical implementation, many functional verification teams abandon the RTL model in favor of a single gates level model. Alternatively, other verification teams will check for equality by running regression simulations on the gate level model and comparing results against the RTL. The problem with these approaches is that they lack a clear separation of circuit equality verification vs. circuit functionality verification. For all but the smallest designs, using simulation to prove equivalence is incomplete. In addition, both approaches severely impact a design's time-to-market metric since gate-level simulation is required, which as discussed in Chapter 4 runs one to two orders of magnitude slower than RTL simulation. The methodology we are recommending requires that the verification process (e.g., either simulation or model checking) remain at the RT-level throughout the duration of the project. To promote this methodology, the formal verification process of *equivalence checking* must be used to completely validate equality on all design transformations.

Consistency. In addition to the challenges of proving equivalence, increasingly complex designs have revealed increasingly complex problems. For years, the system design community has observed a class of functional verification problems, which is inherently difficult to identify either through traditional simulation, emulation testing, or even under a post-silicon design verification lab environment. To define these problems requires understanding their *byzantine* nature--which can be characterized as possessing a complex set of interactions between multiple components or processors, and requiring a unique (and usually long) sequence of events to demonstrate the failure. Unfortunately, the probability of generating the unique and complex sequence of interactions necessary to expose these problems is unlikely under random test simulations, and usually too intricately involved to anticipate using traditional directed simulation techniques.

To address this class of verification problems, researchers have been studying state space-exploration techniques (and other mathematical approaches) to prove correctness. These approaches at first glance appear to offer a solution to all of today's complex functional verification problems. In reality, expectations must be properly set when applying state space exploration techniques, particularly at the RT-level of design. In Chapter 2 we indicated that system, algorithmic or architectural level characteristics are more effectively validated on a higher-level model (e.g., an executable specification written in C, C++, or SDL). Similarly, successful use of state-space exploration techniques to prove chip or algorithmic level properties will typically require a higher-level and more abstract model than is represented by the RTL model.

To successfully apply model checking to the RT-level requires partitioning the large design into smaller verifiable blocks, and then creating a valid environment description for each partition (e.g., a testbench or set of constraints to model the block-level environment). The interface-based design approach discussed in Chapter 2 can simplify the partitioning and modeling effort. It is important, however, not to underestimate the effort required to construct and debug the testbench or constraint-driven environment during the formal verification process.

There are still many interesting RT-level (i.e., generally lower level, not system level) properties and potential corner case concerns that can benefit from the use of model checking. For example, in Chapter 2 we discussed the controllability and observability challenges with simulation and coverage on internal blocks of the design. State-space exploration techniques applied to these internal blocks will provide exhaustive coverage on "hard-to-be-sure" properties. The challenge of applying formal functional verification at the RT-level is limiting the set of model checking properties to only those with the highest coverage concern. This is due to the cost and effort required to model the block-level environment. Examples of successfully applying RT-level model checking include: queue controller underflow or overflow conditions, error correction encode and decode circuits, bus contention, one-hot state machines, block-level interacting state machines, etc.

From our experience, even with all of the limitations and complexities associated with RT-level model checking, the formal verification process can provide the following overall design and verification benefits:

- discover *high-quality* (complex) bugs
- reveal space/performance improvement opportunities during verification
- serve as a prototyping exploratory tool
- contribute to a more in-depth design review
- serve as a "second pair of eyes" for simulation
- increased understanding, confidence and quality of design

In this chapter, we introduce the notion of a finite state machine and its analysis and applicability to proving machine equivalence and FSM properties through state-space exploration. We then separate our discussion of the *RT-level* formal verification process into *transformation verification* (e.g., equivalence checking) and *functional verification* (e.g., model checking). A broad-based discussion on RTL constraints, properties, coding styles and methodologies are presented, which we have found improves the overall equivalence and RT-level model checking process. Finally, we illustrate how to use the assertion checkers (introduced in Chapter 2 for simulation) can contribute to an improved RT-level formal functional verification process. The use of assertion checkers removes formal verification language (and tool) details from the design engineer and allows the engineer to identify RT-level corner case concerns that need verification attention.

5.1 Formal Verification Introduction

Formal verification is a systematic method of ensuring that a design's *implementation* matches its *specification*. In a traditional verification flow, a Design Verification (DV) engineer develops a set of tests, based on his interpretation of the design's specification, coupled with his understanding of the design's implementation. Design correctness is then established through simulated results. Unfortunately, this approach provides no assurance that all corner cases have been covered.

One naive approach to *formal verification* is to enumerate all possible cases, then run directed tests to cover each case. Although this approach is theoretically possible, it is combinatorially intractable for all but the smallest designs.

Formal verification techniques have been developed that use a mathematical proof, rather than simulation and test vectors, to provide a higher level of verification confidence on certain properties [Clarke and Kurshan 1997] [Clarke and Wing 1996]. For example, the *implementation* can be either a Verilog RTL module or an abstract version of the design, while the *specification* is a set of properties (i.e. expected behavior) to be verified, expressed in a suitable form. The proof will then show a relationship between the implementation and specification, and, without test vectors, provides a complete verification for each specification property under consideration (i.e. corner cases are completely covered for the specified property).

The most mature class of RTL formal verification tools is known *as equivalence checkers* [Huang and Cheng 1998]. This class of tools mathematically proves the logical equivalence between different refinements of a design without simulation and test vectors. Specifically, equivalence checkers are used for implementation verification and are created to answer the question: *"Has my implementation been preserved during process transformations?"* Equivalence checkers, in general, are easily integrated into existing design flows with minimal changes. Unlike simulation, they provide a fast and complete verification of equivalence for a given set of constraints (e.g., a gate-level netlist versus its RTL functional model).

Model checking is another class of formal verification tools [McMillan 1993] [Kurshan 1994]. These tools answer the question: *"Does my implementation satisfy the properties of my specification?"* Fairness within a system's bus arbitration is a classic example of a design specification property that can be verified against the design's implementation using a model checker (e.g., a bus access will always eventually be granted to a client requesting the bus).

Formal tools and methods, however, are not a single-point (or complete) solution to today's system design verification problems. All verification methodologies (e.g., simulation, equivalence and model checking) still suffer the consequence of potentially masking real problems by over constraining a design's input.

5.2 Finite State Machines

The process of traversing (or exploring) a finite state machine's state-space is fundamental to understanding formal verification techniques. Intuitively, state-space exploration is the most appropriate coverage metric for identifying bugs. For example, the byzantine class of problems previously discussed are generally due to a unique and complex interaction between finite state machines (FSM), which in theory should be uncovered during state-space exploration. In reality, successful state-space traversal is limited to properties directly involving approximately 200 state bits. To generate an appreciation for state-space exploration, this section introduces a definition for an FSM and its analysis and applicability when proving (a) machine equivalence and (b) FSM static and temporal properties.

[Figure. 5.1] illustrates a Huffman model representation of an FSM. This model consist of the m-tuple input variables $\{x_0, x_1, x_2, \ldots x_{m-1}\}$, the n-tuple output variables $\{z_0, z_1, z_2 \ldots z_{n-1}\}$, the p-tuple of current state element variables $\{q_0, q_1, q_2, \ldots q_{p-1}\}$, and the next-state variables $\{q_0', q_1', q_2', \ldots q_{p-1}'\}$.

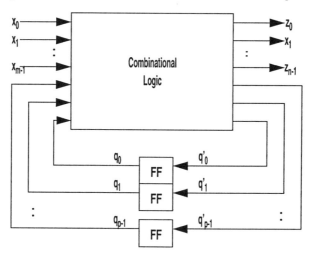

Figure 5-1 Huffman FSM Model

Formally, a *finite state machine* is denoted as a 6-tuple $M=(X,Z,S,s_0,\delta,\lambda)$, where:

- X represents the machine's input space, which is a set of m-tuple input vectors $\xi_i = \{x_0, x_1, x_2, \ldots x_{m-1}\} \in X$ (e.g., $X_0 = (0,0,0,\ldots 0)$).

- Z represents the machine's output space, which is a set of n-tuple output vectors $\zeta_j = \{z_0, z_1, z_2, \ldots z_{n-1}\} \in Z$ (e.g., $\zeta_0 = (0,0,0,\ldots 0)$).

- S represents the machine's reachable state space, which consisting of the p-tuple set of state element variables $s_k = \{q_0, q_1, q_2, \ldots q_{p-1}\}$. The finite state machine has a maximum of 2^p possible states; however, not all states are necessarily reachable. Hence, $S \subseteq \{s_0, s_1, s_2, \ldots s_{2^{p-1}}\}$.

- s_0 is the initial reset state.

- δ is a set of mapping functions from the present state of the machine to the next state based on the values of an input vectors $\xi_i \in X$ $(0 \leq i < 2^m)$. In other words, $\delta: X \times S \rightarrow S$. For example, $q'_j = \delta_j(\xi_i, s_k)$, where $(0 \leq j < p)$ and $s_k \in S$ for $(0 \leq k < 2^{p-1})$.

- λ is a set of mapping functions from the present state of the machine to an output variable z_l $(0 \leq l < p)$ based on the values of an input vector $\xi_i \in X$ $(0 \leq i < 2^m)$. In other words, $\lambda: X \times S \rightarrow Z$.

To traverse the FSM state space, the sets of reachable states are iteratively calculated by a process known as *image computation*. In other words, by starting with the FSM's initial reset state $S_0 = \{s_0\}$, a new set of reachable states $S_1 = \{s_0, \ldots s_k\}$ is calculated by applying the set of δ mapping functions to all state elements for all input vectors $\xi_i \in X$ $(0 \leq i < 2^m)$. This process continues for $S_0 \subset S_1 \subset \ldots \subset S_i$. When the newest set of reachable states is identical to the previous set of reachable states (e.g., $S_{i+1} = S_i$), the iterative process terminates. This is known as a *fixed-point calculation*, as shown in [Figure 5-2]

A breakthrough in FSM state space traversal occurred in the late 1980s when researchers began representing state transition relationships *implicitly* rather than representing explicit state enumeration by state transition graphs or tables [McMillan 1993]. For example, the transition relation for machine M is defined as $T: X \times S \times S$. Specifically, $(\xi_i, s_j, s_k) \in T$ iff state s_j transitions to s_k for the input vector $\xi_i \in X$ $(0 \leq i < 2^m)$. Symbolic techniques (e.g., BDDs [Bryant 1986]) are commonly used to efficiently represent the set of transition relationships. Even with the breakthroughs in FSM state space representation, there are many designs whose state space is too large to represent implicitly, resulting in a condition commonly known as *memory state explosion*.

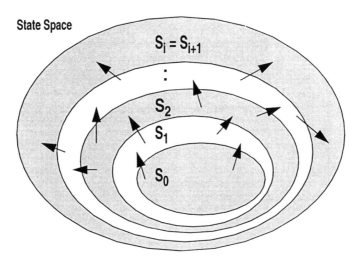

Figure 5-2 Fixed-Point Image Computation

5.2.1 Machine Equivalence

[Figure 5-3] represents a simple model, known as a *miter*, that is used for proving equivalence. This model is constructed by XORing the primary output pairs for the two machines forming what is referred to as a *product machine*. The product machine is a tautology '0' for any input vector $\xi_i \in X$ ($0 \leq i < 2^m$) and any reachable state $s_k \in S$ ($0 \leq k < 2^p$) when the specification machine is equivalent to the implementation machine. Otherwise, the two machines are not equivalent, and the input vector ξ_I and reachable state s_k form a unique *distinguishing vector* used to demonstrate the inequality.The problem presented when proving equivalence can be simplified by maintaining a consistent state space and state encoding between the specification (e.g., RTL) and the implementation (e.g., gate-level netlist). Thus, a process intensive reachability analysis (i.e. image computation of S) is no longer required by the equivalence checking tool. Additionally, the equivalence check problem is reduced to proving combinational equivalence.

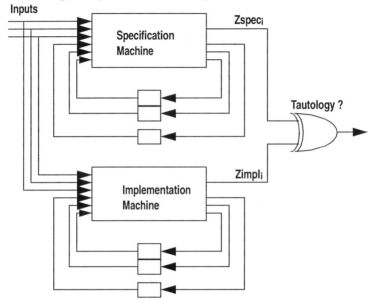

Figure 5-3 Simple miter model for proving equivalence.

For *combinational equivalence checking* [Matsunaga 1996][Burch and Singhal 1998] (e.g., two machines with a consistent state space, state encoding and reset state), state elements are first identified and then mapped between designs. In effect, the state-elements themselves can be abstracted away from the equivalence checking process. Thus, the p-tuple of current state element variables $\{q_0, q_1, q_2, \ldots q_{p-1}\}$ are now viewed as circuit inputs and added to the input set X.

Likewise, the next-state variables $\{q_0', q_1', q_2', \ldots q_{p-1}'\}$ are viewed as circuit outputs and added to the set Z, and the δ set of mapping functions is added to the set of out mapping functions. As a result, the λ set of mapping functions are now only dependent on the values of the input vector $\xi_i = \{x_0, x_1, x_2, \ldots x_{m-1}, q_0, q_1, q_2, \ldots q_{p-1}\}$, where $\xi_i \in X$. ($0 \leq i < 2^{m+p}$). Hence, $\lambda: X \to Z$ and equivalence is now determined by the tautology defined in [Equation 5-1] for the output variable z_l, where ($0 \leq l < 2^{n+p}$):[1]

Equation 5-1: $\lambda spec_l(\xi_i) \oplus \lambda impl_l(\xi_i) = 0$

---

1. Equation 5-1 can also be written to account for the don't care input space $D(X)$. The tautology would then appear as $(\lambda spec_l(\xi_i) \oplus \lambda impl_l(\xi_i)) \mid D(\xi_i) = 0$, where $\xi_i \in X$. Accounting for the don't care space can be eliminated, provided that all case statements within the RTL are fully specified as recommended and the use of X is eliminated from the RTL as recommended in Chapters 4, 6 and 7.

5.2.2 FSM Property Verification

Invariance and *liveness* are common properties often proved on FSMs. *Invariant* or *safety* properties are valid for all time (e.g., some good event should always occur during a given state, or we assert that some bad event should never occur).[2] Invariant checking can be efficiently performed during the process of image computation, and the sequence of states (e.g., an error trace) leading to the failing property is easily generated during reachability analysis.

Liveness properties on an FSM are valid at some future point in time (e.g., eventually some event should happen). For temporal logic model checkers, these properties require a step of *preimage computation* in addition to reachability analysis. Formally, we define the *preimage* of set S as: $\text{Preimage}(S) = \{s_j \mid \exists \xi_i \in X, \exists s_k \in S, \text{ and } \exists (\xi_i, s_j, s_k) \in T\}$. In other words, the preimage is a set of states whose transition relation for a given input vector ξ_i will lead from state s_j to s_k, where $s_k \in S$.

5.3 Formal Transformation Verification

5.3.1 Equivalence Checking

The following sections describe the processes involved in a typical design flow and the process points at which equivalence checking can be applied (See [Figure 5-4]).

5.3.1.1 Equivalence Checking Flow

RTL Refinement . In the initial stages of the design flow, engineers occasionally tweak their RTL coding style in an attempt to improve simulation performance, or explore different coding styles for an improved synthesis mapping. Equivalence checking at this step ensures that the original functionality of the design has been preserved during the RTL refinement.

RTL `ifdef Implementation . Central to our Verifiable RTL methodology is the concept that the RTL remain the main or golden model throughout the course of design. Hence, our functional verification process can focus its

---

2. For a complete discussion of events and assertions, see Chapter 2.

effort on a faster RTL model as opposed to a slower gate-level model.

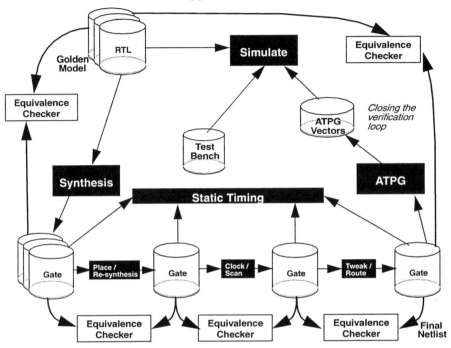

Figure 5-4 Equivalence Checking Flow

Frequently, to secure the timing performance of logic synthesis, design engineers need to specify explicit vendor macro cell instances within the RTL module specification. To prevent the loss of the designer's functional intent, and to prevent the RTL code from degenerating into a slower-simulating gate-level netlist, we recommend coding the RTL as illustrated in [Example 5-1]:

Example 5-1:

```
`ifdef  IMPLEMENTATION
    <macro cell instance implementation>
`else
    <RTL behavior specification>
`endif
```

For example:

```
`ifdef IMPLEMENTATION
    wire t1, t2, t3;
    XOR3 u1 (t1, in[0], in[1], in[2]);
    XOR3 u2 (t2, in[3], in[4], in[5]);
    XOR3 u3 (t3, in[6], in[7], in[8]);
    XOR3 u4 (perr, t1, t2, t3);
`else
    assign perr = ^in; // calculate parity on 'in'
`endif
```

We define the IMPLEMENTATION text macro during the synthesis process and undefine it during simulation. This combination preserves clarity in the RTL description and optimizes the RTL for simulation performance, while insuring a specific implementation during synthesis. At this point, we use equivalence checking to verify consistency between the RTL behavior description and the macro cell instance implementation--simply a self-compare on the RTL module. Similarly, during the equivalence checker's compilation process, we define the IMPLEMENTATION text macro for the implementation, and undefine it for the specification model.

Logic Synthesis. Synthesis is a process of mapping the RTL specification to a gate level implementation. Since simulations should remain exclusively RTL, ensuring that the original RTL specification is correctly mapped to its gate implementation is accomplished using equivalence checking.

Floor Planning. Floor planning is a placement process of major blocks of logic. In many design flows, the physical hierarchy of the design is initially used to match the logical hierarchy of the design. During the floor plan editing process, however, the physical hierarchy of the design can be changed to anything the engineer desires. Equivalence checking at this step ensures the original functionality of the design has been preserved after floor plan editing.

Placement. Placement is a process of fixing each floor planned macro cell instance to a unique location on the chip die with the goal of minimizing wire routing interconnect between each macro cell. The netlist would not normally change during the placement process. Equivalence checking, however, can be used at this step to ensure that the original functionality of the design has been preserved.

Layout and Placement Based Optimization. Layout and placement based optimization (LBO and PBO) [Bening *et. al.* 1997] can be thought of as an elaborate in-place optimization step. This optimization step is used to reduce logic, build and place optimal fanout trees, and change macro cell power levels (i.e. transistor sizes) to the minimum necessary to meet timing. Equivalence checking must be used at this step to ensure that the original functionality of the design has been preserved.

Clock Tree and Scan Chain Insertion. Clock tree insertion is a step of synthesizing an ASIC's clock distribution network. Scan chain hookup is a process of stitching together a set of registers into a scannable ring of sequential elements. To prove equivalence requires that a constraint be applied to disable the implemented scan logic paths when comparing against an RTL model without a scan chain.

Routing. Routing is the process of specifying the specific layers of metal and the exact routing interconnect to be used between the various macro cells. Equivalence checking at this step ensures that all editing to the wiring interconnect has preserved the original functionality of the design.

Timing Tweak Editor. The tweak editor is used to quickly fix *last minute* timing breaks or late (yet simple) functional RTL changes by interactively editing the netlist and placement. This is the point in the physical flow at which most logic errors are inadvertently inserted into the design. Equivalence checking is critical at this step to ensure the RTL specification matches the new netlist and placement.

Vendor Interface / Final Netlist. The final design is translated into the chip vendor's proprietary format. A final Verilog netlist is generated from this translated design and equivalence checking must be used to verify that the final netlist implementation matches its original RTL specification.

5.3.1.2 Closing the Verification Loop

[Figure 5-4] provides a high level view of a recommended equivalence checking flow. Two essential elements of this flow are the final full chip hierarchical-RTL to flat-gate equivalence check, and a simulation of the implemented gate's ATPG vectors on the RTL specification. These two processes effectively *close the verification loop* by identifying any revision-control or process flow errors.

Intuitively, if the design engineer is rigorously checking the design after each transformation, everything should be correct. Experience has revealed that any late updates to source or globally shared include files in the RTL specification, which were inadvertently omitted from synthesis, could be missed without a final chip RTL-to-gate equivalence check. As another example, many processes within the physical design flow involve looping. Any missed verification while looping on a specific process point along the normal design flow would invalidate the flow's transitive relationship of equivalence. Closing the verification loop will detect these errors.

In general, if a few small changes are made to the design during each process transformation, it is easier to find comparison points during the equivalence check process. The sum of these transformation deltas, however, could make identifying these points extremely difficult. Again, our experience has shown that ensuring the various design transformations apply the **Retain Useful Information Principle** (see Chapter 2), as well as the design methodology embracing the **Object-Oriented Hardware Design Principle**, sufficient comparison points are identified to complete the equivalency proof for the entire design.

Simulating the ATPG vectors on the RTL specification can be effectively accomplished provided: (a) all case statements within the RTL are fully specified as recommended in Chapter 6, (b) the scan chain is back annotated into the RTL design as proposed by the OOHD methodology in Chapter 3. *Closing the verification loop* provides a final validation on all libraries used during the synthesis, place-and-route and ATPG processes. These errors might otherwise be missed when equivalence checking at various sequential processes along a design flow (i.e. not referencing back to the original RTL specification). In addition, *closing the verification loop* on the RTL golden model validates the synthesis tool (and the equivalence checker) by identifying most RTL coding style or interpretation differences between the simulator and the synthesis tool.

5.3.2 Cutpoint Definition

The use of internal signal pairs was first proposed by Berman and Trevillyan [1989] to reduce a larger equivalence checking problem into a process of checking a set of smaller related functions. These internal signal equivalent pairs, referred to as *cutpoints*, form the boundaries for corresponding functions between a specification and its implementation. In [Figure 5-5], we illustrate the concept of partitioning a large *cone of logic* into a set of smaller cones of logic, which can be proved independently.

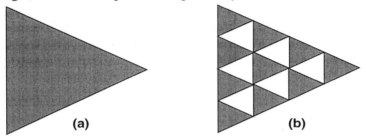

(a)   **(b)**

Figure 5-5 Cutpoint partitioning. (a)Large cone of logic. (b) Set of partitioned cones.

In [Example 5-2], we show an example of a simple RTL specification:

Example 5-2:

```
assign   c_a = f1;
assign   c_b = f2;
assign   c_c = f3;
assign   c_y =(c_a & c_b) | c_c;
```

In [Figure 5-6], we show an equivalent gate implementation to the specification described in [Example 5-2]:

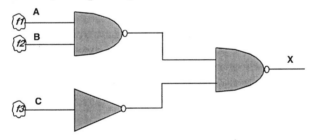

Figure 5-6 Gate Implementation

The RTL specification variables c_a, c_b, c_c, and c_y can be mapped into the gate implementation's functionally corresponding points as shown in [Example 5-3]:

Example 5-3:

Map (c_a, A)
Map (c_b, B)
Map (c_c, C)
Map (c_y, X)

In general, *mapping points* consist of a set of corresponding design specification and implementation latches, input and output ports and internal signal equivalent pairs or *cutpoints*. Many approaches have been proposed and implemented to facilitate the identification of these corresponding points [Brand 1993] [Berman and Trevillyan 1989] [Cerny and Mauras 1990] [Foster 1998] [Kuehlmann and Krohm 1997] [Kunz 1993].

5.3.3 Equivalence Checking RTL Coding Style

In this sections, we examine RTL coding styles that can impact the performance of the equivalence checking tool.

5.3.3.1 Isolating Functional Complexity

Cutpoints were defined in the previous section as a mechanism to speed up the equivalence checking verification process. In general, the designer's RTL should be coded so that the equivalence checking tool can efficiently exploit the structural similarities of the RTL specification and the gate-level implementation.

Cutpoint Identification Principle

A single design decision pertaining to functional complexity must be isolated and localized within a module to facilitate equivalence checking cutpoint identification.

Isolating the functional complexity of multipliers is a classic example of how the equivalence checker's runtime performance can be improved by applying the *Cutpoint Identification Principle*.

In [Example 5-4], a 16x16 multiplier has been combined with other sub-expression to form a complex expression.

Example 5-4

assign c_indx = (((coord_x * coord_y) &
                  indx_mask) + indx_offset);

In [Example 5-5], the functional complexity of the multiplier has been isolated within a module instantiation. The equivalence checking process will benefit by the clean interface when exploring the structural similarities between the specification and the implementation. In addition, instantiating the mult_16x16 module permits referencing a process specific optimized library. For example, simulation can potentially take advantage of the host machine multiplication while modeling the multiplier. Similarly, the synthesis or equivalence checking process can take advantage of a uniquely specified multiplication algorithm (e.g., Booth), optimized specifically for the appropriate process.

Example 5-5

mult_16x16 mult1 (coord_x, coor_y, mult1_prod);
assign c_indx = ((mult1_prod **&** indx_mask) + indx_offset);

5.3.3.2 Test Expressions within Case Statements

Complex *test expressions* within a Verilog *case* statement can complicate the verification process. It is easier to debug the branching effect within a simulation trace file when the case test expression is an observable variable. In addition, equivalence checking the RTL specification to a gate level implementation is improved through potentially additional observability or cutpoints.

Test Expression Observability Principle

A complex test expression within a case *or* if *statement must be factored into a variable assignment.*

[Example 5-6] provides an illustration of functional complexity within the *test expression* of *case* statements.

Example 5-6

case ((a **&** b | c ^ d) || mem[idx])
  4'**b**0100: c_nxt_st = r_nxt_st << 1;
  4'**b**1000: c_nxt_st = r_nxt_st >> 1;
  default: c_nxt_st = r_nxt_st;
endcase;

In [Example 5-7], we show the test expression after factoring the complex expression into a variable assignment.

Example 5-7

c_nxt_st_test = (a **&** b | c ^ d) || mem[idx];
case (c_nxt_st_test)
  4'**b**0100: c_nxt_st = r_nxt_st << 1;
  4'**b**1000: c_nxt_st = r_nxt_st >> 1;
  default: c_nxt_st = r_nxt_st;
endcase;

By adapting the Test Expression Observability Principle, the resulting RTL is easier to debug during simulation and provides potential cutpoints during equivalence checking.

5.3.3.3 Equivalence Checking OOHD Practices

Chapter 3 introduced the practices of object-oriented hardware design. This section will show how these practices are applied to optimize the equivalence checking process.

A large percentage of the equivalence-checking process time is expended while identifying register or latch pair mapping points. The identification of these mapping points and, in general, the entire equivalence-checking process can benefit from an OOHD practice and tool-specific library methodology

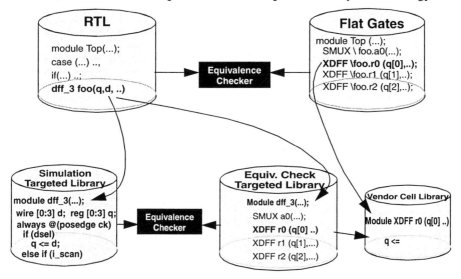

Figure 5-7 Hierarchical-to-Flat name mapping.

Register Name Mapping. The OOHD methodology provides an equivalence checking optimization by automatically identifying all register and latch pairs between the RTL specification and its gate-level implementation. For example, by referencing the equivalence checking or synthesis targeted library we are able to directly map the hierarchical RTL instances to their synthesized netlist instances.[3] [Figure 5-7] illustrates the method of referencing the synthesis targeted library from the RTL during the equivalence-checking process. Note that the hierarchical RTL reference 'foo.r0.q' is directly mapped into the

---

3. The RTL synthesis-target library in our example flow contains actual vendor macro cells for the flip-flops and input muxes (i.e. we control all synthesis of registers and their related input muxes by library references). See reference [Bening *et al.* 1997] for details on this physical flow using text macros.

flat gate-level netlist reference '\foo.r0 .q' due to the synthesis (or equivalence checking) targeted library referencing the shared vendor cell library. The equivalence-checking targeted library is validated against the higher-level simulation targeted library efficiently by using an equivalence checker.

Master-Slave Latch Pair Folding. Another equivalence checking optimization advantage the OOHD methodology provides is a mapping a two-state master-slave latch pair to a single state-point representation (within the RTL description). This is accomplished by referencing an equivalence-checking optimized targeted library for the state-element object. The equivalence-checking targeted library is constructed to accurately model the master-slave functionality for each bit, within the functional grouping of state elements instantiated in the RTL (e.g., a 16 bit instantiated register can be optimally modeled for equivalence-checking using 16 lower-level master-slave latch pairs). Without utilizing an equivalence-checking targeted library, the equivalence checker is forced to invoke routines to identify the two-to-one state mapping between the implementation and the specification.

5.4 Formal Functional Verification

This section describes: (a) techniques for capturing design specification properties and assumptions directly within the RTL code, (b) methods of coding to prevent state explosion and facilitate an efficient model checking process, (c) adapting OOHD practices to solve multi-phase clock abstracting and seamless support for multiple tool coding requirements, and (d) pre- and post-silicon model checking methodologies.

5.4.1 Specification

The successful integration of block level RTL model checking into the design flow require unambiguous specifications. Specifically, what design block level output properties are required for verification? During verification, what block-level environmental assumptions must be put to use? Are the input assumptions valid?

In Chapter 2, we discuss specification as a process of creating a precise and unambiguous description of the design's behavior. We emphasized that it is the act of specification that enables the engineer to acquire an intimate understanding of the design space and ultimately uncover design deficiencies prior to RTL coding. Furthermore, the product of specification provides an effective device useful for both communication and analysis.

At the time of this publication, there is no Open Verilog International

(OVI) standard for capturing a design's specification within the RTL. A number of commercial tools have developed their own proprietary languages or meta-languages, which are embedded into the RTL as comments and can be used to describe a design's properties and assumptions. Likewise, a few public domain tools permit embedded temporal property languages (such as CTL [Clarke, Emerson and Sistla 1986]) directly into the Verilog RTL as meta-comments. Even with the lack of a standard RTL specification language, successful RTL block level model checking can be achieved when the design's expected behavior and assumed operating environment have been specified.

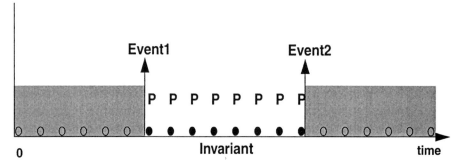

Figure 5-8 Invariant Property Window

As described in Chapter 2, most temporal relationships can be viewed as an event triggered window, bounding an event or assertion. For example, [Figure 5-8] illustrates an *invariant* property. The assertion P, in this example, is valid after *Event 1* occurs, and remains valid until *Event 2*. The event-trigger can be expressed using any Verilog expression, or expressed as a specific point in time. Likewise, the assertion P can be expressed using any Verilog expression. For an invariant, the assertion P would be qualified with either *always* or *never,* indicating its occurrence within the event bounded window.

5.4.1.1 Properties

[Example 5-8] illustrates a *RTL specification pseudo-language* [4] that could be used to describe the event-triggered window shown in [Figure 5-8].

---

4. The *RTL specification pseudo-language* described in this chapter is based on a subset proposed by Kurshan [1997]. Our intentions are not to propose a new specification language. Our intentions are, however, to provide a simple educational tool that can be used to illustrate many, but not all, RT-level formal verification property and constraint concepts. The *RTL specification pseudo-language* is deliberately model checker neutral.

Example 5-8

PROPERTY QueueSafe // *for output port queue_underflow*
   GIVEN *(queue_valid==1)*
   ASSERT NEVER *(queue_underflow==1)*
   UNTIL *(queue_valid==0)*

This specification states the following: given that the event-trigger *(queue_valid==1)* occurs, the claim (or assertion) is that *(queue_underflow==1)* will never occur, unless the assertion occurs after the event-trigger *(queue_valid==0)*.

Alternatively, this property could be captured in the RTL using a CTL meta-comment as shown in [Example 5-9]

Example 5-9

//assert: QueueSafe: **AG** ((queue_valid==1)) $\Rightarrow$
                                  (queue_underflow!=1) **U** (queue_valid==0)))

The embedded CTL property is read directly from (or extracted out of) the RTL code and used by many public domain model checkers, such as *SMV* [McMillan 1993] or *VIS* [Brayton et al. 1996]. Note that the RTL specification pseudo-language described in this chapter is deliberately model checker neutral. Our goal for developing the pseudo-language was to provide a simple mechanism for capturing RT-level properties directly into the RTL source. This was necessary due to the lack of an Open Verilog International (OVI) formal property language standard. For our methodology, properties captured using this pseudo-language are automatically translated into the many different commercial and public domain model checking tool property languages using a simple *perl* script. Capturing properties (or test targets) directly in the RTL preserves the designers concern for potential corner cases that might require extra verification attention. This is another example of applying the Retain Useful Information Principle.

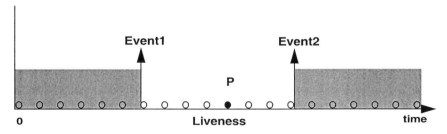

Figure 5-9 Liveness Property Window

As another example, [Figure 5-9] illustrates a *liveness* property. The assumption *P,* in this example, will eventually occur after the first event-trigger occurs and before the occurrence of the second event-trigger.

The RTL specification pseudo-language for [Figure 5-9] is illustrated in [Example 5-10]. This specification states: given that the event-trigger *(reset & request)* occurs, the claim (or assertion) is that *(grant==1)* will eventually happen within a specific time unit of eight clocks after event-trigger 1 (e.g., the time at given event-trigger--*tg*).

Example 5-10

PROPERTY CntrlLive *// Controller Liveness Property*
    GIVEN *(reset & request)* *// Event-trigger 1*
    ASSERT EVENTUALLY *(grant==1)*
    UNTIL *[t=tg+8]* *// Event-trigger 2*

Multiple *properties* can be combine to specify complex behavior using the connectives &, |. For example, P1 & P2 | P3.

5.4.1.2 Constraints

Capturing the design's input assumptions (or assumed input behavior) during specification and design is as important as capturing the design's output or internal properties. During the process of logic simulation, equivalence checking, and model checking, the design assumptions (also known as *constraints*) are used to restrict or constrain the input space behavior.

In general, constraints can be classified as *static* or *temporal.* [Example 5-11] demonstrates a static constraint. In this example, the reset_ signal is restricted to its inactive value throughout the verification process. Since the static constraint (i.e. our assertion) is not bounded by any event-triggers, the GIVEN and UNTIL constructs are not required.

Example 5-11

ASSUMPTION Reset *// Example of a static constraint*
    ASSERT ALWAYS *(reset_=1)*

Temporal constraint behavior can be specified using a combination of assumptions and the propositional connectives & and |. In addition, [Example 5-12], demonstrates how non-determinism can be modeled using the EVENTUALLY construct:

Example 5-12

ASSUMPTION ReqND *// Non-deterministic input request*
   GIVEN *(request==0)&& (grant==0) at [t=0]*
   ASSERT EVENTUALLY *(request==1)*
   UNTIL *(grant==1)*

5.4.1.3 RTL Specification Pseudo-Language

This section defines a pseudo-language that could be used to specify the behavior for block-level RTL. In addition, a method for embedding the specification pseudo-language into the RTL is described.

The following BNF provides a description for the RTL simplified specification pseudo-language:

{PROPERTY I ASSUMPTION} *<Name>*
   [GIVEN { *<Event> I <Verilog expr> [at <time spec>]*}**]**
   ASSERT {EVENTUALLY I ALWAYS I NEVER I AT *<time spec>*} *<Verilog expr>*
   [UNTIL *<Event>*]

Where an <Event> can be defined either as a <Verilog expr> (i.e. Verilog expression) or as a specific <time spec> (e.g., at time zero [t=0], or 8 clocks after the initial (or GIVEN) event-trigger [t=tg+8], etc.).

As previously stated, complex properties and assumptions can be formulated by using a combination of assumptions and the propositional connectives & and I during the verification process.

In [Example 5-13], the `ifdef SPECIFICATION construct embeds the specification pseudo language directly into the RTL code. This is analogous to the `ifdef IMPLEMENTATION construct described for equivalence checking in section 5.3.1.1. One advantage of using the `ifdef construct is that it permits the creation of non-synthesized logic or state machines for trapping events. Note that for this simple property could have been proved without the extra state machine used to trap the event. The extra state machine was added to this example to illustrate the concept of creating a trap that could be used for more complicated properties.

Example 5-13

```
`ifdef SPECIFICATION // for formal verification pre-processor
// Design Assumptions (i.e. constraints) and Properties
        ASSUMPTION ResetLow // Non-deterministic reset
                GIVEN (reset_==1'b0) at [t=0]
                ASSERT EVENTUALLY (reset_==1'b1)
        ASSUMPTION ResetHigh
                GIVEN (reset_==1'b1)
                ASSERT NEVER (reset_==1'b0)
        ASSUMPTION Init1
                GIVEN (init==1'b0) at [t=0]
                ASSERT EVENTUALLY (init==1'b1)
        ASSUMPTION Init0
                GIVEN (init==1'b1)
                ASSERT EVENTUALLY (init==1'b0)
        PROPERTY SequenceLive
                GIVEN (reset==1'b1) && (init==1'b0)
                ASSERT EVENTUALLY (found_sequence==1'b1)
                UNTIL (init==1'b1)
// Simple (not synthesized) state machine used to trap an event for verification.
        reg state, found_sequence;
        initial found_sequence=1'b0;
        initial state=1'b0;
        always @(posedge ck) begin
          case (state)
          0: if ((reset_==1'b1) && (init==1'b0) begin
             found_sequence <= 1'b0;
             state <= 1'b1;
             end
          1: if ((reset_==1'b0) || init==1'b1))
             state <= 1'b0;
             else if (start_sequence==1'b1)
             found_sequence <= 1'b1;
          endcase;
             end
       `endif
```

5.4.2 Model Checking and Parameterized Modules

Queue structures, in general, must be treated as independent objects and abstracted away from other functionality within the Verilog RTL. In other words, a parameterized model for the queue should be instantiated to provide a mechanism for queue size reduction during verification. Queue depth and word size reduction will result, in many instances, in an improvement in the model checking runtime performance while potentially preventing the condition of state explosion. This is achieved by applying the *Numeric Value Parameterization Principle.*

Numeric Value Parameterization Principle

Numeric values should be parameterized and not directly hard-coded into the RTL source.

5.4.3 Model Checking OOHD Practices

This section describes the advantages of adopting the OOHD practices, which were described in Chapter 3, when integrating model checking into a design flow. It is important to note that the OOHD methodology does not optimize the typical problems associated with model checking (i.e. state explosion). This methodology, however, enables the verification engineer to overcome various tool specific coding restrictions and the ability to abstract out phase-related clocks across multiple clock domains without disturbing the engineer's original RTL source. Overcoming tool coding and usage restrictions in a seamless and unobtrusive step within the design flow is an essential point for adopting an OOHD methodology.

Coding Restrictions and Initialization: In our experience, a few public domain model checker translators require that all registers be coded using the Verilog blocking assignment as opposed to the typical register non-blocking assignment. Without requiring the design engineer to re-code their RTL for tool evaluation or integration, the OOHD methodology permits the generation of a model-checker targeted library with the optimal tool-specific coding policy. In addition, the targeted library enables the designer to overcome some of the public domain model checkers inability to globally initialize state elements by including initial blocks directly within the library model. It is true that there are commercial tools available without these coding restrictions and initialization problems. The point we would like to emphasize, however, is that the OOHD methodology permits its users to evaluate multiple tools (with and without coding and usage restrictions) quickly and in a seamless fashion.

Clock Abstraction: The second coding restriction the OOHD methodology permits us to solve deals with modifying all state-element clocking descriptions to support a common clock within the model checking tool. This is problematic when there are different phase-related versions of a common clock distributed throughout the design (e.g., a common clock *ck* and a half speed clock *ck_hs*). Without clock abstraction, many designs will encounter memory explosion during state-space traversal since the BDD data structures can effectively double when accounting for the clock.

To support *multi-phase related clock abstraction* during the model-checking process, the OOHD methodology permits distributing a phase enable signal directly to all state elements in the design during the model checking process and then uses a single common clock. For example, the OOHD pre-processor (described in Chapter 3) will automatically identify all phase related half speed clock signals (e.g., ck_hs). Then the pre-processor replaces the phase related clock with the common clock (e.g., ck), and creates a new interface port with an appropriate phase enable signal. The model-checking targeted library will now describe the state element, using the phase enable signal, as follows in [Example 5-14]:

Example 5-14

```
module dff_xc_4 ( q, d, scan_in, ck, scan_sel, reset_, phase_en);
    output  [3:0] q;
    input  [3:0] d, scan_in;
    input ck, scan_sel, reset_, phase_en;
    reg [3:0] q;
    always @(posedge ck) begin // model checking model
        if (phase_en) // to support common clock abstraction
            q <= (reset_ == 0) ? 4'b0000 : ((scan_sel) ? scan_in : d);
    end
endmodule
```

For designs containing multiple phase-related clocks, the OOHD design abstraction and tool-specific library methodology provides a technique for abstracting out the clock during the model-checking process.

5.5 Model Checking Methodology

5.5.1 Pre-Silicon Model Checking Methodology

In Chapter 2, we introduced the use of event monitors and assertion checkers as a mechanism for capturing and validating design assumptions and properties during simulation. We showed how an assertion-targeted library could be generated and used during the simulation-debugging phase. This library contains runtime checks that will halt the simulation process upon the detection of an error. Furthermore, an assertion-targeted library can be generated for the faster simulating regression-phase by providing a mechanism for logging events, which will be post-processed and analyzed for correctness.

Alternatively, an assertion-targeted library can be generated containing our RTL specification pseudo-language (or a proprietary tool's specification language) and then used within a formal verification flow. As an example, the QueueSafe property in [Example 5-8] could be captured in the RTL using the assert_never assertion checker defined in Chapter 2 (see [Example 5-15]):

Example 5-15

```
assert_never queue_safe (ck, (queue_valid==1'b1),
                             (queue_underflow==1'b0),
                             (queue_valid==1'b0),
                             'ASSERT_SAFE_1);
```

This assertion can either be used by the simulation or formal verification process. The assertion-targeted library would define the assert_never module as shown in [Example 5-16]:

Example 5-16

```
'define DELAY_ASSERT #2;
module assert_never (ck, event_trig_1, test, event_trig_2, serverity_level);
input ck, event_trig_1, test, event_trig_2;
input [7:0] serverity_level;

'ifdef SPECIFICATION // for formal verification pre-processor
   PROPERTY assert_never
      GIVEN event_trig_1
      ASSERT NEVER test
      UNTIL event_trig_2
'endif

'ifdef ASSERTION_CHECKER_ON // for simulation
   reg test_state;
   initial test_state=1'b0;
   always @(event_trig_1 or event_trig_2)
      if (event_trig_2 || event_trig_1)
         test_state = (~event_trig_2) && (event_trig_1 || test_state);
   always @(posedge ck) begin
      'DELAY_ASSERT
      if((test_state==1'b1) && (test==1'b1)) begin
         $display("ASSERTION ERROR %d:%t:%m", serverity_level, $time);
         $finish;
      end
   end
'endif
endmodule
```

By encapsulating the assertion checking code within a module, the engineer can concentrate on the specific properties of the design he wants verified--leaving the details of the checking mechanism to the verification group. This provides the verification group with the ability to optimize and tune the assertion-targeted library throughout the duration of a project, without interfering with the text or functional intent of the original RTL description. It enables the formal verification engineer to create various model checking assertion-targeted libraries written in a specific commercial proprietary specification language (or our neutral RTL specification pseudo-language that can be automatically translated to a specific model checking tool). See Chapter 2 and 8 for additional details on an assertion-targeted library.

5.5.2 Post-Silicon Model Checking Methodology

For many situations, it is possible to query a model checker for a specific lower level property, which has been found failing in the lab (e.g., a queue underflow condition). A benefit from reproducing the lab bug in a model checker is the creation of a formal testbench, which can then be reused to verify the revised design--covering all corner cases associated with the original failing property.

It is important to understand, however, that even though we are able to formally verify the original failing property on the revised design, this does not mean all properties of the design are verified. In other words, the revised circuit might interact with other portions of the design to create a new bug. Hence, it is still necessary to run regression simulations to complete our verification of the entire design.

5.6 Summary

In the late 1980's, the system design community underwent tremendous productivity gains in gate-level design due to engineers embracing synthesis technology. Unfortunately, this resulted in an increase in the design verification problem space. To keep up with escalating design complexity and sizes, we recommend that design verification engineers augment their traditional verification flows with *formal methods*. This chapter introduced concepts and tools that are used in a formal verification design flow with an emphasis on RTL coding styles and methods of specification that lend themselves to efficient equivalence and model checking.

A question quit often asked by design verification managers is "do formal verification tools really increase design productivity?" Our answer is "it depends." Certainly formal equivalence checking has enabled us to replace regression simulation--achieving greater verification coverage in minimal time. The benefit of using this technology has been a reduction in development time-to-market--while simultaneously achieving a higher level of verification confidence in the final product. However, in an industrial environment the RT-level model checking process has not fared as well as equivalence checking. One of the biggest challenge facing the successful integration of model checking into today's design flow is not entirely due to technology and tool limitations. Just as significant is (a) the reluctance of engineers to adopt a design discipline that cooperates with the formal process, (b) the dearth of usable and unambiguous design specifications. Primarily, it is the act of specification that enables us to achieve an intrinsic understanding of the design space and ultimately uncover design deficiencies prior to the process of design and verification. Ultimately, it will be the act of formalism that will increase our design productivity by assisting both the traditional and formal verification process and tools.

In this chapter, we introduced the notion of a finite state machine and its analysis and applicability to proving machine equivalence and FSM properties. We then separated our discussion of the formal verification process into *transformation verification* (e.g., equivalence checking) and *functional verification* (e.g., model checking). In addition, we also discussed coding styles and methodologies which will improve the overall equivalence and model checking process. Finally, we illustrated how the assertion checkers, described in Chapter 2 for simulation, could be leveraged during the formal functional verification process.

6

Verifiable RTL
Style

Writing RTL Verilog following a verifiable style serves both EDA tool success and the productivity of the other engineers who participate in the design process.

The verifiable RTL Verilog style ideas in this chapter draw from the authors' experiences developing and supporting an RTL-based design flow on large system (> 200 million gate-equivalents) projects. We have found other good sources of ideas in:

- prior writings on reuse [Keating and Bricaud 1999].
- gems from the internet comp.lang.verilog news group.

To prior writings on Verilog style, this chapter adds additional principles of style based on verification, as well as explanations of the reasoning behind the style.

As well as general style elements for major modules, this chapter also prescribes specific style elements applicable to smaller (around 20 lines or less), high usage modules from a library. The special style rules for the small modules in libraries are listed in sections 6.2.1.4 and 6.4.

6.1 Design Content

6.1.1 Asynchronous Logic

With the ever increasing clock speeds of synchronous logic domains, the asynchronous boundaries between these domains are closer together than ever before. The need to deal with resynchronization within the box enclosing a single large system is becoming widespread.

RTL simulation can perform some sampled test relationships of asynchronous transactions, but there are two other tools which are both more important than RTL simulation for asynchronous verification. Details about these tools are beyond the scope of this book, but they are important to mention so that the reader is appropriately oriented and can know where to look for more information.

Asynchronous protocol verification. Telecommunications system designers have had to verify asynchronous communications protocols for decades, and have generally turned to Petri net models. Murata [1989] presents an excellent survey of Petri net modeling.

Petri net modeling requires mapping an asynchronous protocol design into an abstract graph. If the design is tractable for Petri net model reachability analysis, it is possible to know whether asynchronous events in different sequences can lead to adverse states (deadlock, livelock, etc.) If the design turns out to be intractable, it must be simplified.

Resynchronization failure rate analysis. All mechanisms for resynchronization between asynchronous time domains have a metastable failure rate. For a given resynchronization frequency rate and resync circuit speed, it is possible to calculate a mean time between failure (MTBF) rate. Safe design practices considering resynchronization frequency and circuit speeds can result in predicted failure rate of one every ten billion years or more. Kleeman and Cantoni [1987] present an excellent explanation of metastable failure rate design considerations.

It is of utmost importance for design project members to know that an increased resynchronization frequency rate or a slower resync circuit speed can result in a catastrophic change in failure rates. Failure rates can go from one every million years to three per day by doubling the frequency.

Asynchronous Principle

A design project must minimize and isolate resynchronization logic between asynchronous clock domains.

Asynchronous protocol verification and resynchronization analysis are rare engineering specialties, which makes them expensive.

In addition to costs of asynchronous protocol and failure analysis, there is the case derived from testability for minimizing and isolating resynchronization logic between asynchronous clock domains. The theory and methods for test are not well-developed across asynchronous boundaries.

6.1.2 Combinational Feedback

In modern logic design practice, combinational logic feedback is universally avoided. Verification tools that count on no combinational logic feedback (cycle-based simulators, boolean equivalence checkers, timing verifiers) diagnose such feedback loops.

Though rare, designers sometimes inadvertently specify combinational feedback loops in their RTL. Three sources of feedback loops that can hinder RTL verification flow are design errors, false paths and apparent (not real) feedback.

Design error. These are real unintended combinational feedback loops. About half the time they will have an odd number of inversions in the feedback path. The resultant inversion can result in short or 0-delay oscillations in an event-driven simulator. Before the availability of tools that would identify feedback loops in RTL Verilog, 0-delay oscillations in a simulation would take several frustrating days of a designer's time to locate and fix.

False path. Every other year or so, a clever designer will come up with a way to correct a short-path timing problem by cross-coupling combinational logic in a manner that can never be fully enabled. [Figure 6-1] shows an example of two cross-coupled multiplexers with an inverted select signal on one of the multiplexers. Note that with the select line s at a logic 0, the feedback path from y is disabled, and with s at a logic 1, the feedback path from z is disabled.

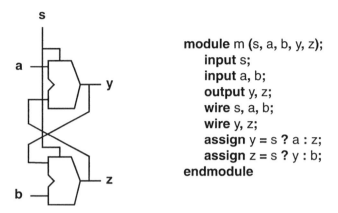

Figure 6-1. False-path feedback.

Apparent. The combinational feedback checking performed by verification tools that are sensitive to feedback is often fast and simple. Simple checking may identify a combinational logic feedback condition where none actually exists. [Example 6-1] (a) illustrates some Verilog code that can appear to have combinational logic feedback. A simple feedback checker treats the always procedural block as a black box, with all inputs connected to all outputs.

Example 6-1

a) Module with apparent feedback	**b)** Inlining assign to clear up feedback	**c)** Moving procedural assignments to assign

```
module m (a, d);
    input a;
    output d;
    reg b, d;
    wire c;
    always @(a or c)
        begin
            b = a;
            d = c;
        end
    assign c = b;
endmodule
```

```
module m (a, d);
    input a;
    output d;
    reg b, c, d;
    always @(a)
        begin
            b = a;
            c = b;
            d = c;
        end
endmodule
```

```
module m (a, d);
    input a;
    output d;
    wire b, c, d;
    assign b = a;
    assign d = c;
    assign c = b;
endmodule
```

[Example 6-1] (b) shows one remedy in which the designer moves the **assign** inside the procedural block in functionally correct evaluation order, and with no apparent feedback.

[Example 6-1] (c) shows another apparent feedback remedy in which the designer moves the procedural assignments into independent **assign** statements. Use of **assign** statement relieves the designer of the burden of ordering the statements, since verification tools that need to rank-order for their function take care of rank-ordering in their compilation process. However, the verification labor costs will be greater if the evaluation order of combinational logic is not immediately apparent to a reader.

Combinational Feedback Principle

Designers must not use any form of combinational logic feedback (real, false-path, apparent) in their Verilog.

6.1.3 Case Statements

In Chapter 4, we pointed out how case statements provide a significant part of the simulation performance advantage of RTL simulation over gate-level simulation. We also showed there how two-state simulation at the RT-level results in better verification of case statements (and if-else statements) than simulating with an X-state.

In Chapter 5, we discussed how test *signals* in case statements support formal verification better than test *expressions* in case statements.

In this section, we discuss two case statement style requirements that support design verification processes: fully-specified case statements and case test signal and constant consistency. Throughout this section, the case statement style requirements apply to **casex** as well as **case** statements.

6.1.3.1 Fully-Specified case Statements

Verifiable RTL design requires that designers fully specify case statements. Designers must specify all resultant output values corresponding to 2^N input values, where N is the bit-width of the case statement control signal. [Example 6-2] illustrates a **case** statement that specifies next states for only the intended zero-to-two three-state counter function (a), and for all values of r_o, including the "impossible" 2'**b**11.

Example 6-2
a) Functionally-specified case   b) Fully-specified case

```
module c (r_o, c_n);                module c (r_o, c_n);
    input [1:0] r_o;                     input [1:0] r_o;
    output [1:0] c_n;                    output [1:0] c_n;
    reg [1:0] c_n;                       reg [1:0] c_n;
    always @(r_o)                        always @(r_o)
    case (r_o) // rtl_synthesis full_case    case (r_o)
        2'b00 : c_n = 2'b01;                 2'b00 : c_n = 2'b01;
        2'b01 : c_n = 2'b10;                 2'b01 : c_n = 2'b10;
        2'b10 : c_n = 2'b00;                 2'b10 : c_n = 2'b00;
    endcase                                  2'b11 : c_n = 2'b00;
endmodule                                endcase
                                     endmodule
```

To many designers in many design groups, use of fully-specified case statements throughout their entire system design is a major departure from their general practice in prior work. To these designers, the way to handle the impossible states in a **case** statement is to:

1. add a default and assign an "x" to the next-state signal, and

2. issue a diagnostic message.

Because the authors cannot assume that very many of the readers are apriori in sympathy with use of fully-specified case statements throughout a system, we now present some justification.

First, the use of fully-specified case statements for an entire design not a theoretical idea that the authors have only used in experiments. It is based on its application in an actual large-scale design project comprised of millions of gates on chips, 100's of these chips in systems, and thousands of these systems delivered to customers [Bening *et al.* 1997].

Next, we know that there are advantages and disadvantages to fully-specifying case statements. Even though the authors feel that verifiability tips the balance in favor of fully-specified case statements, the following itemizations list advantages and disadvantages that the reader should be aware of.

The verification advantages that come from fully specifying case statements are:

• Boolean equivalence checking performance.

Not having to deal with the exponential number of don't care sets in boolean equivalence results in a typical 5-10X boolean equivalence checking performance improvement, and makes it possible to prove equivalence on some cones that would otherwise be unprovable in any practical amount of run time.

• RTL - gate-level simulation alignment.

When the RTL specifies a logic output for all values, designers can be certain that the RTL simulation model precisely corresponds cycle-by-cycle and state-by-state to what would happen in the gate-level simulation model.

• Improved RTL simulation performance/verification

Completely specified case statements eliminate the need for an X-state, improving RTL simulation performance, as well as startup state testing, as described by one of the authors in [Bening 1999b].

• RTL manufacturing test simulation.

The RTL - gate-level simulation alignment provided by fully-specified case statements allows a design project to run its manufacturing test vector simulations against the RTL. This feature provides a 5-10X performance advantage in simulation run times compared to gate-level simulation, as well as an overall double check on boolean equivalence, gate-level test generation process, and the overall design flow.

Verification considerations with respect to fully-specifying case statements run opposite to other design considerations which have weighed in favor of partial specification. These include synthesis minimization, test generation, and loss of simplicity.

• Synthesis minimization.

We have done synthesis experiments using the 8-bit one-hot decoder module shown in [Example 6-3]. Fully-specifying the case statement with the **default** included, a synthesis tool generated 3X as many gates as only specifying the eight one-hot cases without the **default**.

In counter argument, there are other ways to specify logic that results in full-specification while still achieving timing and gate-minimization. An example is presented later in this section. We have found that when creative designers on a project pool their ideas, they can use full specification while still achieving timing and area goals.

Example 6-3

```
module one_hot(c_hot,c_code);
    input [7:0] c_hot;
    output [2:0] c_code;
    reg [2:0] c_code;
    always @ (c_hot) begin
      case (c_hot) // RTL synthesis full_case
        8'b10000000: c_code = 3'b000;
        8'b01000000: c_code = 3'b001;
        8'b00100000: c_code = 3'b010;
        8'b00010000: c_code = 3'b011;
        8'b00001000: c_code = 3'b100;
        8'b00000100: c_code = 3'b101;
        8'b00000010: c_code = 3'b110;
        8'b00000001: c_code = 3'b111;
                default: c_code = 3'b000; // causes 3X gates
      endcase
    end // always (c_hot)
endmodule // one_hot
```

- Test generation.

  Combining two separately synthesized blocks in which one block generates a one-hot encoded signal and the other block receives the signal using
  a fully-specified case statement will result in redundant logic. Redundancies present test generation methods with logic that may result in untest-
  able sections [Abramovici *et al* 1990 pp. 100-103].

  In counter argument, one-hot encoding is just one source of many possible
  redundancies introduced between separately synthesized logic blocks, as
  designers combine them prior to test generation.

- Loss of simplicity.

  Throughout this book, we argue for simplicity as a fundamental ingredient
  of RTL-based verification. However, to express a design using fully-specified case statements while still achieving sufficient gate minimization to
  achieve area and timing goals often requires that a designer code Verilog
  using more statements.

  In counter argument, the descriptive methods for fully-specifying logic
  while achieving timing and area goals form common statement patterns in
  the Verilog for a design project, and soon become easily recognized (and
  welcome) by verification-oriented engineers.

  Productive designers draw from examples, from the Verilog that they have
  written in previous projects, and from synthesis guidelines in manuals. Most

readers will likely find plenty of incompletely specified case statements from their sources of examples. To counter this situation, we now present examples of fully-specified case statements for various functionalities.

Small case statements. For case statements that use a control variable which is only two or three bits wide, the increased gate cost of fully-specifying all possibly states is small. [Example 6-2] showed a small case statement where fully-specifying all values of the control variable does not add many gates.

One-hot encoding. [Example 6-4] illustrates a way to process the one-hot encoding shown in [Example 6-3] to get faster gates. It tests each bit to get the next state, and separately test for the error condition illustrates one-hot decoding in a fully-specified manner that results in a minimum gate count from synthesis.

Example 6-4

```
module one_hot(c_hot,c_code);
    input [7:0] c_hot;
    output [2:0] c_code;
    reg [2:0] c_code;
    reg [2:0] c_code0,c_code1,c_code2,c_code3;
    reg [2:0] c_code4,c_code5,c_code6;
    always @ (c_hot) begin
      c_code6 = (c_hot [6]) ? 3'b001 : 3'b000;
      c_code5 = (c_hot [5]) ? 3'b010 : 3'b000;
      c_code4 = (c_hot [4]) ? 3'b011 : 3'b000;
      c_code3 = (c_hot [3]) ? 3'b100 : 3'b000;
      c_code2 = (c_hot [2]) ? 3'b101 : 3'b000;
      c_code1 = (c_hot [1]) ? 3'b110 : 3'b000;
      c_code0 = (c_hot [0]) ? 3'b111 : 3'b000;
      c_code = c_code0 | c_code1 | c_code2 | c_code3 |
               c_code4 | c_code5 | c_code6;
    end // always (c_hot)
endmodule // one_hot
```

The test for the error condition can either be implemented in actual hardware (for the situation where the data path carrying the c_hot one-hot signal is not reliable), or as a verification assertion, or both

[Example 6-5] shows how the error condition for the one-hot signal can be detected as a separate **case** statement test. It demonstrates how to isolate the on-hot decode complexities for error-testing. Notice how we separate the correct operation path from the error test. In [Example 6-4], the correct operation path, the logic only does a test of one bit, which makes room for several levels more logic before the next clocked register stage.

[Example 6-5] also illustrates use of encapsulation (see Chapter 2) of error logging treatment details. The assert_one_hot is a non-hardware macro module that addresses verification needs, both simulation and formal methods. Without changing the design Verilog, verification engineers can progressively refine the assert_one_hot with additional internal controls to deal with simulation of:

- good machine testing and error testing. With good machine testing, an engineer would likely want the simulation to stop if the one-hot encoding failed. With error testing, in which the engineer injects errors to see whether the hardware recovery or shut down processes run correctly, the engineer would likely want the simulation to continue so that many errors can be injected and tested in one run.

- pre and post reset simulation. Prior to completion of a reset sequence, the engineer would not want the simulator to stop when detecting an error in one-hot encoding, but after the reset completes, the simulation should log encoding errors

- diagnosis and regression. For the fastest possible regression simulation, reduced logged data is a good idea.

Example 6-5

```
...
case (c_hot)
    8'b00000001,
    8'b00000010,
    8'b00000100,
    8'b00001000,
    8'b00010000,
    8'b00100000,
    8'b01000000,
    8'b10000000: c_hot_error = 1'b0;
    default: c_hot_error = 1'b1;
endcase
end // always (c_hot)
assert_one_hot c_hot_check (c_hot);
endmodule // one_hot
```

And, with a different library definition, assert_one_hot can serve as a formal model checking constraint and property. See Chapter 5 for details.

In this section, we have presented arguments favoring fully-specifying case statements, and one example showing how a designer can achieve timing and area goals while fully-specifying the logic outcomes in the RTL. To develop fully-specified RTL design solutions for all design functions needed on a project would require an entire book devoted to the topic.

We recommend that the project set full-specification as a requirement, and then share the fully-specified design solutions to timing and area problems that their designers develop across the project.

6.1.3.2 Test Signal and Constant Widths

There are two kinds of consistency designers must maintain in the case statements for their designs: signal-to-constant, and constant-to-constant.

Verifiable RTL design requires that the bit width of the case test signal match the bit width of the constants with which it compares. Simulators do not check for this, but lint-type rule checkers do perform the check. This is fortunate, since such a mismatch invariably means a design oversight that results in an eventual simulation test failure, which in turn results in computer time and human labor costs.

In **case** statements, each constant *must* be unique. Duplicate constants always mean an oversight by the designer, and they can often cause bugs that become hard to diagnose. We have seen situations where a designer adds a case constant and some action, then wonders why the action never happens, not noticing that there was an identical constant value earlier in the case statement with another action. As with width conflicts, case constant duplication often results in an eventual simulation test failure, with similar costs and frustration.

In **casex** statements, each constant *should* be unique, or specify a unique range if it contains don't-care "?" values. The ranges should not overlap, or encompass one another. We weaken the checking for **casex** statements compared to **case** because some designers assert that there are a few logic functionalities that are expressed much more efficiently with overlap.

To perform constant duplication and overlap checks, lint rule check implementors have independently arrived at a bit-mapping algorithm that counts on the width of the case test signal and the constants being less than 31 bits. Limiting case test signal width to less that 32 bits also speeds simulation by allowing a direct conversion of the Verilog case statement to simulation host machine instructions that support programming language case statements. (This assumes a 32-bit simulation host machine word length).

6.1.4 Tri-State Buses

Tri-state buses present specialized challenges to design and verification. Timing verification, boolean equivalence, and logic simulation are tools that need to treat tri-state buses as exceptions to their normal boolean logic processing. The object-oriented RTL design methodology basics introduced in Chapter 3 points the way towards encapsulation and selection of a uniform practice for specifying tri-state function.

Drivers. [Example 6-6] illustrates the encapsulation and uniform practice for specifying a tri-state driver.

Example 6-6

```verilog
module dff_tri_4(pin,qin,ck,dis);
output [3:0] pin;
input [3:0] qin;
input ck;
input dis;
reg [3:0] qout;
always @(posedge ck)
    qout <= qin;
tri [3:0] pin = !dis ? qout : 4'bZ;
endmodule
```

The expression sequence across the tri statement must be uniform in all tri-state drivers in a design: test, normal output and Z output.

In tri-state buses, designers often need pull-down (or pull-up) resistors to keep CMOS transistors from creating a short-circuit path and burning out in an adverse power-up combination of states. To express the start up state function of these resistors for simulation at the register transfer level, use **tri0** and **tr1** in place of **tri**. Specify the net driven by a **tri, tri0** or **tr1** up and down the hierarchy as type **wire** wherever it connects up and down the hierarchy. The net function "inherits" the pull-down/pull-up function from the driver declaration.

Receivers. Encapsulation of tri-state bus receivers within library modules localizes the special treatment of the boundary between:

- a net that can be at a Z state and
- the boolean two-state logic.

In RTL verification, there are cycle-by-cycle state-by-state design errors that need to be addressed:

- multiple drivers on a tri-state bus.

- a receiver that is active when there are no drivers on a tri-state bus.

Both of these are best addressed by assertions, but the assertions can be supplemented by simulation modeling style that intercepts Z's (or X's in the case of bus multi-driver conflicts). shows a tri-state receiver in which the user task $InitialState plugs random 0/1 bit values into the holding signal qpin. Then, every time a new value comes in on pin, the user task $Trapxz plugs in the random 0/1 bits from qpin into ipin corresponding to the bit positions in pin that are Z (or X).

Example 6-7
module rec_tri_4(pin);
inout [3:0] pin;
reg [3:0] ipin,qpin;
always @(pin **or** qpin) **begin**
    ipin = pin;
    $Trapxz(ipin,pin,qpin);
end
initial
    $InitialState(qpin);
endmodule

6.2 Organization

6.2.1 System Organization

6.2.1.1 Compiler Options

Defining constants. In Verilog, it is good system design practice to specify constants by name within sub-block modules, and assign their value using **'define** to specify constant value assignments.

Good places for named constants are case statements specifying state machines, and bit-widths/depths of reusable data path modules, such as fifo buffers, register files and memories.

For constant names shared across multiple modules, use **'include** "*file-name*" within each module, and specify the constant definitions only once in a single file.

Compared to **'define**, parameter-based constants defined by **parameter** and **defparam** statements may at first appear to be the better choice for specifying constant values. They are better behaved when a design project integrates sub-block modules into system blocks in that they retain their definition only within the modules containing their definitions. The fact that **parameter** and **defparam** definitions end at the **endmodule** statement provides a cleaner mechanism for controlling the scope of constant definitions.

However, parameters have shortcomings as listed below.

- The per-instance constant substitution that gives parameters a power that `define` constant definitions don't have is provided with greater flexibility by the preprocessor-based techniques described in Chapter 3.

- Parameters can introduce simulation run time penalties in cases where a designer uses run time parameter values to check for a instance configuration that is constant-based.

- Parameters have a higher degree of difficulty than `define` in their implementation by EDA tool developers. As well as startups, major established vendors have differences (bugs?) in their implementations of parameters.

We recommend that you let your competitors spend their time on the parameter issues while your project sails on to success without parameters.

In Verilog language parsing, `define` macro definitions remain "alive" from the point in the Verilog text stream input where they are defined, independent of module and file boundaries. To avoid macro name clashes when integrating their system model, a design project must have a macro definition naming convention.

Code Inclusion Control. The `ifdef - `else - `endif conditional compilation directives can control inclusion of test assertions, diagnostic statements, waveform output PLI calls, and RTL abstractions.

For the fastest possible logic simulation later in a design project when design bugs are very rare, reducing diagnostic aids in the Verilog source is desirable. If one simulation out of 10,000 fails, having to rerun the failed simulation with the diagnostic aids included is a good trade-off when the 9,999 simulations run 2X faster without the diagnostic aids.

Even though inclusion control could be done by `define` statements within the Verilog input files, it is more flexible to do the control by EDA tool command line macro definition options. These are generally of the form

    ... +define+INITSTATE+RECORDOFF+ ...

Code inclusion control supports productive simulation-based verification, where all design verification simulations must be done using RTL abstraction for best simulation performance. In some design blocks, synthesis and data-path mapping from the RTL cannot produce a gate-level design that meets the design constraints. For these blocks, boolean equivalence must be used to prove that the RTL version of the design exactly matches another version that is closer to gate-level. [Example 6-8] shows the inclusion control statements bracketing gate-level and RTL versions of the same logic, and the corresponding boolean equivalence command line options.

Example 6-8

a) Verilog Source

```
    ...
'ifdef IMPLEMENTATION
            gate-level version of logic
    'else
            RTL version of logic
    'endif
    ...
```

b) Boolean equivalence command line options

-model1 a.v -model2 +define+IMPLEMENTATION a.v

Command line **+define+** code inclusion controls are global to the all the Verilog files that an EDA tool's Verilog compiler reads. On large design projects, the Verilog files for a design are the work product of a large number of engineers, so there is a danger that code inclusion controls may become complicated and redundant if their definition and use are not coordinated.

Code Inclusion Control Principle

A design project must define and document code inclusion controls and provide a process for managing them.

6.2.1.2 Design Hierarchy

For design of computer systems, design projects divide up their design descriptions into a hierarchy of modules.

The top-level modules specify interconnect above the chip-level, usually PC boards and multi-chip modules. For verifying a system design, we believe that the hierarchy of modules describing the instances of chips, daughterboards and motherboards must match the hardware hierarchy.

Within a large chip, the first level down from the chip module consists of a clock tapper module and a core logic module, as shown in [Example 6-9]. The tapper module takes the clock and test controls as input, and generates special clocks (half-speed, phased) as needed, plus test outputs. The core logic module instantiates the modules which perform the function of the chip.

Example 6-9

module chip0(
    chip timing inputs,
    chip function i-o's);
    tapper1 tapper(
        chip timing inputs
        core timing outputs);
    chip_core1 core(
        chip function i-o's);
endmodule

module chip_tapper1(	**module** chip_core1(
chip timin inputs,	*core timing inputs,*
core timing inputs);	*chip function i-o's*);
timing control procedures	*chip major module instances*
endmodule	**endmodule**

This design hierarchy method isolates the complication of clock timing and test from the function of the chip. Clock and test specialists can focus their design and verification effort on the tapper, and leave the chip core designs to focus on the chip's function.

Starting early in the project, design engineers can lint the design from the core module on down to ensure that its part of the design is fully compliant with cycle-based simulation. Then, later in the project, when simulation performance is a greater concern, the core part of the chip design can be simulated with all of the performance advantage of cycle-based techniques.

6.2.1.3 Files

File names. Efficient team-based design and verification requires that a separate file hold a single module description, and that the file name be the same as the module name. Because separate small files come up faster in editors, it is natural for design engineers to break up their Verilog for a large design into separate files.

However, some designers may think that it is a good idea to put a set of smaller modules that have a strong functional relationship into a single file. The grouping of modules into a file may help the design engineer, but it will inevitably cost more than it saves. Other engineers verifying the design and running it through the EDA tool flow will be repeatedly be hindered in their work when trying to find a module that is buried within a file. Compared to issues of design size and complexity, the hindrance is not a big deal, but it does add some cost that the originating design engineer must consider.

The only exception to the one-module-per-file rule is shared and standard modules stored together in a library file.

Files in directory hierarchy. We have seen some designers closely map their module design hierarchy into a directory hierarchy, such that a module file that is five levels down in the design hierarchy is also five levels down in the directory hierarchy.

The better way is to have some grouping of design module levels at each directory level. This practice provides some knowledge of functionally-related sets of modules at a glance.

Note that the use of a design level suffix digit on the module/file name is very helpful to the engineer who is picking up someone else's design. Here we have an example of six hierarchically suffixed file in one directory.

```
pa_cam_loc7.v     pa_resp_dec6.v     pa_resp_in4.v
pa_resp_arb5.v    pa_resp_find6.v    pa_resp_side5.v
```

Modern hierarchical browsers have reduced the importance of the file organization in a directory hierarchy, but not eliminated it. Some unifying scheme of file and directory structure greatly simplifies the task of setting up the list of files for verification tools, which includes hierarchical browsers.

6.2.1.4 Libraries

Library files contain all of the encapsulated and reusable modules for a design. These modules include all flip-flops, memories, input-output/tri-state drivers and receivers, as well as other design modules that fit encapsulation or reuse goals. Examples of the other design modules that may fit into a library include clock-generators, parity trees, multiplexers, error-correcting encode/decode logic, and queues.

In addition to providing the mechanism for sharing of module designs across the project, they also support the object-oriented concepts described in Chapter 3.

In our design work, we have typically used two to four libraries, where each library contains the module definitions related to a chip design technology (fabrication, test), and grouped by function. Flip-flops and muxes would be in one library, and memories in another.

Here are some organizing methods within a library:

- Include a comment header for the entire library file, containing a complete body of header information as described in section 6.5.2.1.

- Group modules with similar function, e.g., flip-flop modules followed by input-output modules.

- Place comments in each library module in proportion to the complexity of the module. For modules about fifteen lines or less, a minimum is a one-line comment spelling out the function of the module in words.

  // D-flip-flop with scan, reset, and inverted output
  module dff_srn (...

- From the convention established for Verilog gate-level primitives, declare the output of simple Verilog library modules first in the I-O list on the module line. If there is an inverted output, declare it next.

6.2.2 Module Organization

6.2.2.1 Overall Organization

By applying common organization methods to all modules, a project improves the productivity of everyone working on the project, including the originator of the design. Every engineer can know where to look for statement types of interest in their analysis. Moreover, planned **'define** names in project-wide **'include** files common to every design module avoids integration clashes.

In upper-level modules in the design hierarchy, it is good design practice to only instantiate and interconnect major submodules. All behavioral details must be in the lower-level modules.

[Example 6-10] specifies the sequence of statements for upper level modules that instantiate and interconnect major sub-modules in a design.

Example 6-10

```
<header comment block>
'include "<file name>"
module <prefix><name><level number> (input-output);
<input declarations>
    <clocks>
    <reset>
    <data>
<output declarations>
<inout declarations>
<wire declarations>
<major submodule instances - explicit port connections>
endmodule
```

In [Example 6-11], lower-level modules include behavioral statements (**always, assign**) and library module instances in their organization. Note that the library module instances are at the bottom of the module. This allows for the way that the preprocessor (see Chapters 2 and 3) expands module instance macros and adds lines that change line numbers between its input file and its Verilog output file. Many types of diagnostic messages from verification tools that read the Verilog point to line numbers in the Verilog. Putting the module instance macros at the end keeps the line numbers the same for all of the lines preceding them in both the preprocessor input and its generated Verilog file.

Example 6-11

```
<header comment block>
'include "<file name>"
module <prefix><name><level number> (input-output);
<input declarations>
    <clocks>
    <reset>
    <data>
<output declarations>
<inout declarations>
<wire declarations>
<reg declarations>
<always procedural blocks>
<assign statements>
<major submodule instances - explicit port connections>
<library module instances - may have implicit port connections>
endmodule
```

The line numbers for the statements from the beginning of the file and through the **assign** and major module instances are the same between the source and the Verilog output. This makes changing the source file based on EDA tool diagnostic messages about the Verilog line numbers much simpler.

6.2.2.2 Connections

Generally, we believe in explicitly specified port-signal connections. An exception can be made for connecting instantiated library elements in preprocessor source file input. The preprocessor can take the implicitly connected input and generate Verilog with explicit connections, as shown in [Example 6-12]. See Chapter 3 for more details regarding library module instance connection processing.

Example 6-12

a) Implicit port-order based signal connections input

```
DFF (4, reg_head, r_head, ck, c_head);
```

b) Explicit port-signal connections output

```
dff_4 reg_head ( .d(c_head),
                 .ck(ck),
                 .q(r_head));
```

The only kind of expressions allowed on ports are concatenation expressions. Concatenation has special meaning in terms of bus joining and bus splitting, as well as being applicable to input, output and bidirectional port connections.

By putting expressions on ports, you lose the direct observability of the result of the expression, even with concatenation expressions. Observability is discussed with regard to case statement control expressions in Chapter 5.

Expressions on ports usually point to a less than desirable partitioning between module function and interconnection. Separating function from connection makes a module easier to understand, as well as tracking in verification through improved observability.

6.2.3 Expression Organization

In addition to making a design more understandable to engineers reading the Verilog source code, the designer originating the Verilog can avoid verification tool pitfalls by writing well-crafted expressions.

The first tool to get through is the lint check, as described in Chapters 2 and 3. The lint check brings strong typing to the Verilog language expressions. (Without expression checking, the Verilog language is not a suitable vehicle for design description on a commercial project).

The tutorial in Chapter 8 presents specific RTL design rules for all of the allowed operators. Here we present some general principles, repeating some of the rules from Chapter 8, as well as basics and formal methods from Chapters 3 and 5.

Precedence. Use parenthesis instead of operator precedence to get the expression behavior that you want, and document the behavior for engineers reading your Verilog.

Logical operators. Use logical **&&**, **||** and **!** in place of their bit-wise **&**, **|** and ~ only where they fit into expressions. The basic rule is that the **&&**, **||** and **!** apply to one-bit operators. Since the **&**, **|** and ~ apply to one-bit operands as well, a more narrow application of the logical operators in order. The narrow application is in comparison sub-expressions, as in the following code:

```
if (((r_tm_hdr_len == 2'h1) && (r_tm_data_len == 4'h0)) ||
    ((r_tm_hdr_len == 2'h0) && (r_tm_data_len == 4'h1))) ...
```

and to one-bit operands regarded as boolean, as in the following code.

```
if (c_res_ready && (r_req_last || (!c_req_ready))) begin ...
```

Observability. If complex expressions become too long, break them up into two (or more) expressions with an intermediate signal to carry the subexpression value to the next expression. At about four lines for a single complex expression is a good threshold at which the designer should begin thinking about breaking an expression into separate statements.

In addition to providing the observability that is important to formal verification discussed in Chapter 5, adding intermediate signals makes the Verilog easier to debug in simulation.

When an expression consists of many symmetrical subexpressions that are nearly the same from one line to the next, it is all right to run across many more than four lines. About a half page (or 30 lines) is a good threshold for breaking symmetrical subexpressions into separate statements.

Simulation performance. Where there are opportunities, write expressions in a style that improves simulation performance, particularly in high usage library modules. Here are a couple of common expression patterns that occur in many designs.

- Concatenation - In general, Verilog logic simulators perform better with concatenation than with subrange assignments. Instead of:

```
c_x [23:16] = r_a;
c_x [15:8] = r_b;
c_x [7:0] = r_c;
```
use
```
c_x = {r_a, r_b, r_c };
```
The Verilog compilers for some newer simulators are able to perform this optimization automatically, but only in cases where the pattern of statements for subrange assignment are complete and local to a single procedural block of code.

- Parallel value operations. Where there is a repeated logic operation across selected bits of a multi-bit signal, you can simulate faster by using a mask constant to select the bits, then apply the unary operator that matches the logic. This kind of expression happens in content-addressable memories and error-correcting code encodings. Instead of:
  ```
  cam_1 = s[30] | s[26] | s[22] | s[18] | s[14] | s[10] | s[6] | s[2];
  ```
  use
  ```
  cam_1 = | (s & 32'h44444444);
  ```

Note that by a quick inspection, it is easier to verify the correctness of the pattern in the mask than the correctness of the eight subscripts.

While writing Verilog expressions, watch for opportunities to use concatenation and parallel value operations. And, minimize individual bit visits. Even though Verilog compiler writers continually enlarge their set of optimizations, there are plenty of opportunities to code Verilog for simulation performance.

6.3 Naming Conventions

6.3.1 General Considerations

6.3.1.1 Consistency

The main goal of a design language is communication. This includes communication with engineers as well as the programs that aid the engineers in their verification and implementation of a design. In order to facilitate this main goal, design projects must apply consistent naming practices for all of the Verilog that describes a system design.

This consistency begins with use of names that have become widespread and thereby defacto standards throughout the industry. Keating and Bricaud [1999] relate some of these names, like ck and rst. The next step is to add naming conventions to carry the semantics specific to a design project.

Language efficiency is an important consideration when a project devises consistent names for signals, instances, constants, modules, etc.

Note that in all natural languages in the world, high-usage names are short words. Consider *one*, *two*, *three*, *you* and *me* in English and other world languages. It is not an accident that HDL technology has wound up with ck and rst as short names.

Another consistent naming practice for deriving shorter names is throwing out vowels, and some consonants, while still retaining enough letters in a name to at least distinguish it from other nearby names.

6.3.1.2 Upper/Lower Case

A design project can prepare for a smooth flow through the gate and physical design tools by using lower case for ALL **wire**, **reg**, instance, and port names. Mixing the upper/lower case in Verilog designs will work fine through simulation, but will bring schedule and labor costs when putting together the EDA flow for the gate and physical design processes.

Design projects are increasingly using tools that relate the gate-level design back to the RTL behavioral Verilog, like boolean equivalence checking and back-annotation. It is easy to visualize a simple automated full-duplex mapping of all-lower-case names in the RTL behavioral Verilog to all-upper-case names in gate-level design. If the RTL behavioral Verilog uses mixed case for its names, automated full-duplex mapping becomes more than a trivial filter. The labor costs and schedule hits in the EDA flow resulting from arbitrarily mixing upper/lower case are worth avoiding. Following the all-lower-case rule will make every tool that reads your Verilog check for name collisions early in the design process.

Use of upper case characters is the best practice for all constant names and code inclusion control names defined by **'ifdef** and **'define** statements. Upper case clearly distinguishes these names from signals where they are referenced in the in the text. Following this practice, the example,

    if (c_ptr > 'FIFO_DEPTH) ...

instantly shows 'FIFO_DEPTH to be a **'define**, and

    if (c_ptr > fifo_depth) ...

instantly shows fifo_depth to be a **reg** or **wire** signal.

6.3.1.3 Hierarchical Name References

In verifiable RTL design practice, connection by hierarchical name references has two uses. One is to allow an engineer to reference local signal names within the hierarchy of a design from a verification environment, such as a simulation testbench. The other is to specify back annotation of physical-placement-based hookup of a scan chain back into an hierarchical RTL model by means of scan "stitching."

Referencing local signals. The following line illustrates use of a testbench statement to set up configuration values in a design prior to starting a simulation.

> **force** stac0.core.csr.csr_regs.r_toc_config = {2'h1, 1'h0, 1'h1};

The names stac0, core, csr, and csr_regs are module instance names going down the hierarchy, and r_toc_config is a **wire** declared within module instance csr_regs.

Other verification environments in which hierarchical references are important are cutpoints for equivalence checking, and constraints in model checking.

Note how the use of shorter module instance names allows the entire path hierarchy to be expressed on one line. Shorter module instance names help in waveform viewer windows as well, leaving more room for the waveform, while showing the hierarchical path that labels each waveform.

Scan stitching. Simulating scan-based manufacturing test vectors developed from the gates against the RTL chip model benefits a design project in two ways.

1. The test vectors can be simulated far faster (5X or more) at the RT-level than at the gate-level.

2. Running the gate-derived scan vectors against the RTL provides a double-check of synthesis, boolean equivalence tools and gate-level test generation libraries.

Here we have an example of a scan stitch statement consisting of four lines from a 6,747-line file that hierarchically back-annotates RTL scan connections for a small (250K-gate) chip.

```
assign erac0.core.in_p3.pkt_info.r15.scan_in[2:0] = {
            erac0.core.in_p3.pkt_info.r15.q [1],
            erac0.core.out_m1.mux.r2.q [3],
            erac0.core.in_p3.read_1.r2.q [4] };
```

You can see the power of hierarchical references if you consider what it would take to back-annotate scan connections directly into the modules. It would require changes to all of the modules to wire the scan connections through their ports, up and down the hierarchy.

See Chapter 3 for more about the OOHD techniques for making scan stitching work at the RT-level.

6.3.1.4 Global/Local Name Space

As successive projects develop ever-larger chips and systems, global names require an increasing degree of planning in order to prevent name conflicts as the pieces of the design are put together to form the system model. Project-wide allocation of names and unique prefixes for the different major pieces of a design prevents name clashes for global names.

In the Verilog language, the global names that require planning on large design projects include:

- **module** type names
- **'ifdef** - **'else** - **'endif** conditional compilation names
- user task/function names

In addition to the preceding names, entry point names mapped from user task/function names must be unique within the entire design model. Moreover, these user-defined entry-point names must not clash with entry-point names defined by simulation tool vendors in their object-code libraries.

To assure good order in the process of integrating a large system model, planning for the integration is a must.

- **'ifdef** - **'else** - **'endif** conditional compilation names must be allocated from a project-wide data dictionary.
- The project must allocate short prefixes for engineers to use on all of their:
  - module names in their chips
  - user task/function names, and
  - user-defined entry-point names.
- Vendors must use their own uniform sets of prefixes for all of their entry-points on their programming libraries.

It is important to note that the same prefixes that serve to avoid name clashes in the integration of a large system simulation model also help support performance profiling.

6.3.1.5 Profiling Support

With the performance improvements in successive releases of EDA vendor RTL Verilog simulators, an increasing share of the responsibility for poor simulation performance can be attributed to inefficiencies in the way users code their Verilog and PLI C code.

To locate points of simulation inefficiency in the Verilog and PLI C that comprise a system model, design projects turn to performance profiling tools. These tools sample a representative run of a simulation model, and report the percentage of the run time spent in each of functions that comprise the model.

With the ever larger sizes of chip and system simulations, and the related growth in PLI C code, the profiled simulation model may be comprised of 100,000 or more functions. In a profiling report for such a large model, the engineer doing the profile report analysis needs help.

Profiling tools provide some help for grouping related functions where one function calls subordinate functions, and the subordinate functions also call functions, and so on. The help comes from a profiler's ability to sum the time percentage spent in the function itself plus all of the time spent in the subordinate functions and the functions they call. This works well for seeing the performance of a large number of related functions in PLI C code.

However, simulator calls to the code in most Verilog modules as well as some PLI C code functions do not relate to the hierarchy of instances in the design tree. As shown in [Figure 6-2], simulators call blocks of code scattered across the modules and functions from an event manager.

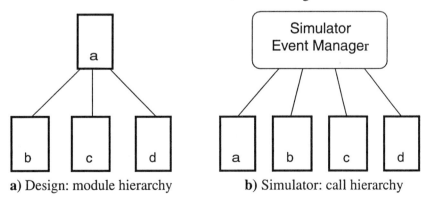

a) Design: module hierarchy   b) Simulator: call hierarchy

Figure 6-2. Design vs. simulator hierarchies.

The profiling report for the simulation of the modules in [Figure 6-2] shows the percentage of host computer time spent in the four modules as independent numbers. If we consider that large system profile reports have 1000's of modules, the absence of information relating submodules to the chip in which they are instantiated makes it difficult to relate performance problems back to the responsible design group.

Profiling support requires that designers prefix their submodule design names with one or two characters that relate all of the submodules to the major block or chip in which they are used. Modules from libraries and shared across chip designs must relate to the library name.

The same holds true for C code in verification tool PLI libraries, as well as simulation and tool vendor libraries. To support profiling of large system simulations, the prefix on all functions within program library must identify the library. Here are some suggested prefixes:

- Simulation user task/function libraries: ut_, ta_, ...
- Simulation and tool vendor libraries: vcs_, nc_, vc_, ...

6.3.2 Specific Naming Conventions

The following subsections itemize naming conventions specific to each named class of Verilog entity. In some cases, a section points back to earlier discussion of background details that support the convention.

6.3.2.1 Constants

Use upper-case names for constants specified by **'define** macro names. See section 6.2.1.1 for more information.

6.3.2.2 File Names

Here are file name suffix-based naming conventions specific to each kind of Verilog file content.

a.s - Contains one **module a() ... endmodule** consisting of the Verilog plus the macro instances for a design. Run this file through a translator to get the following Verilog file for a design block. (See Chapters 2 and 3 for details about the translator and its relationship to the above files.)

a.v - Contains one **module a() ... endmodule** that describes one block in a design. This is the verifiable RTL Verilog version of the design, with the macro instances now in the form of in-line Verilog code or instances of Verilog modules from a library.

file.h - Contains constant definitions in terms of **'define** names for a design. The constants may be state codes or bit ranges.

m.**vlib** - Library of modules.

6.3.2.3 Instances

Instance names can and must be short. Short here means four-to-eight characters typically, and 15 characters as a maximum.

- They can be short because they only need to be unique within a module.
- They must be short in order to keep hierarchical path references from becoming cumbrously long, as discussed in section 6.3.1.3.

Note also that the benefit of short instance names is greater for modules that are above the bottom level of the design hierarchy. These modules are used more often in hierarchical path references, and therefore provide the most benefit in terms of keeping hierarchical path references short. This fits the natural language pattern of higher usage names being short (see 6.3.1.1).

6.3.2.4 Modules

Module names need to be unique across the entire set of modules that comprise a system design. Each chip design might consist of 200 or more modules, so to prevent name conflicts as a project begins combining the Verilog modules for multiple chips into a system, the project must have a module naming methodology.

A simple and effective method is to prefix all of the module names within a chip with the module name for the top-level module in the chip (which is usually the chip name).

Another useful bit of knowledge to carry in the module name is where the module fits into a chip design hierarchy. To do this, we add a number suffix to our module names, where "0" is the top-level module in the chip design, and larger numbers specify lower level modules.

So, given a chip named "ab," the module names within the chip hierarchy would be:

```
module ab0(...);
...
module ab_muxin4(...);
...
module ab_tbd_queue2(...);
...
```

Note that this chip-based prefix naming method can also help improve large multiple-chip system model logic simulation performance, by making simulation performance profiling reports relate far more clearly to the chips that comprise the system. By simply adding up the percentages for all of the functions prefixed with ab_*, the simulation performance analyst can quickly determine whether the ab chip design is important to the overall system model simulation performance. For more information about profiling, see section 6.3.1.5.

As with instance names, engineers more often see and use names for modules that are higher in the design hierarchy, and they therefore can be shorter. This again fits the natural language pattern of higher usage names being short (see 6.5.1.1).

6.3.2.5 Port Names

For verifiable RTL design, consistency in port naming is a must!

Where an output port on one module connects directly to an input port, the port names must be the same. The primary reason for this rule is that it makes it much easier for an engineer reading someone else's Verilog and follow connections through the design.

Another benefit of consistent port naming is that it allows a simple tool to automatically generate connections in the Verilog for the module next up in the hierarchy. [Example 6-13] shows how such a tool does the work of filling in all of the details of connections implied by consistent port names. Space limitations make the example look insignificant, but in actual designs, submodules might have 50 or more ports that a tool can implicitly connect based on consistent port naming.

Example 6-13

a) Submodules with consistent names	**b)** Module instantiating dr5 and rc5	**c)** Consistent names imply connections

module dr5(s, t);
output [1:0] s, t;
endmodule
module rc5(s, t);
input [1:0] s, t;
endmodule
	module nu4();
	dr5 dr();
	rc5 rc();
	endmodule
		module nu4();
		wire [1:0] s, t;
		dr5 dr(.s (s),
		　.t (t));
		rc5 rc(.s (s),
		　.t (t));
		endmodule

Most major Verilog design shops have used ad. hoc. implicit connection tools in-house. Implicit connection tools based on consistent port names are now emerging in the marketplace.

Except for main bus ports and port names connected to signals that fanout everywhere, most port names must be descriptive and somewhat longer than the typical module or instance name. About 10-15 characters is a good rule of thumb for port names.

6.3.2.6 Signal Names

The general rule is that signal names must match port names that they connect, and that they be named after the source signal name or primary input port that drives them.

A signal name policy that matches signal and port names must deal with cases where a module instantiates a submodule two or more times. Common practices are based on suffixing the name carried through the hierarchy, such as the following:

- adding a suffix to the signal name for the connection to each submodule. Here we add an _l and _r to designate left and right.

  wire [15:0] r_bus_l, r_bus_r;

  ...

  r_bus_driver r_bus_driver_l (.r_bus (r_bus_l), ...

  ...

  r_bus_driver r_bus_driver_r (.r_bus (r_bus_r), ...

- using a suffix to indicate a partial bus driver within the submodule. Here we add a _p in the submodule to indicate a partial bus.

  wire [31:01] r_bus;

  ...

  r_bus_driver r_bus_driver_l (.r_bus_p (r_bus[31:16]), ...

  ...

  r_bus_driver r_bus_driver_r (.r_bus_p (r_bus[15:0]), ...

The above examples show only a few cases of what designers encounter when trying to carry consistency up through the hierarchy where multiple instances of a specified module type are involved. Admittedly, in some cases the functionality of a port may change wildly per instance, and the practice of carrying part of the name through the hierarchy becomes impractical.

The name length rule for signals is the same as that for port names: short two-to-eight for high usage names, and up to 15 or so characters for other signal names.

We recommend a signal-naming practice in which signals directly driven by registers have an r_ prefix, and the combinational signals have a c_ prefix. The signal name for the combinational logic cone that drives the input of a register should be the same as the register output except for the c_ and r_ prefix. In [Example 6-14], we have a flip-flip instance with the c_eri_last signal coming in on the data input, and the flip-flop output q driving the registered version of the r_eri_last signal.

Example 6-14

```
dfft_xc reg_eri_last (
        .q(r_eri_last),
        .xq(r_eri_last_n),
        .ck(ck),
        .d(c_eri_last),
        .reset_n(reset_n),
        .i_scan(i_scan)
        );
```

6.3.2.7 User Tasks / Functions and Program Libraries

Like module names, user task and function names relate to a global name space. For the reasons discussed in 6.3.1.4, user task and function naming must consider simulation performance profiling and module integration into a system model.

The authors recommend project-wide allocation of prefixes to user task and function names, as well as their corresponding PLI function library. [Table 6-1] presents some examples of consistent prefix names for all of the entry points for a given library.

Table 6-1. Examples of user task /function calls and library entry points.

User Task Call	Library	Library entry point
$cb_init	libcb.a	cb_init()
$cb_rand	libcb.a	cb_rand()
$xb_datachk	libxb.a	xb_datachk()
$xb_datalog	libxb.a	xb_datalog()

In actual experience with large system models, we have seen as many as 50 libraries, and with around 50 to 100 entry points on each library.

Vendor simulator and simulation tool libraries enter into the global name space and profiling of large system designs. In earlier days, we encountered a name clash between a new release of vendor library and one of our in-house libraries that did not observe our current prefix-based naming method. Performance profiling was encumbered by lack of a coherent naming convention for vendor entry points. We reported these problems to the vendor, and they fixed them by adding a common prefix to their library entry point names.

Entry Point Naming Principle

RTL tool libraries must support a prefix-based entry point naming convention.

Sadly, the entry point names in standard UNIX libraries provide very few unifying conventions in their naming. It is probably a little late to fix.

6.4 Naming In Verilog Library Modules

Because the functionality of library modules is generic and simple, and because they generally have wide usage, their internal port and signal names should be short. [Example 6-15] presents a simulation model that illustrates typical generic short names, like d for data input and q for register output.

Example 6-15

```
// Flip-flop with scan and inverted output
module dff_sn_32(q, q_, ck, d, sc_sel, scan_in);
    output [31:0] q;
    output [31:0] q_;
    input ck;
    input [31:0] d;
    input sc_sel;
    input [31:0] scan_in;
    tri0 [31:0] scan_in;
    reg [31:0] q;
    assign q_ = ~q;
    always @(posedge ck) begin
      q <= sc_sel ? scan_in : d;
    end
endmodule
```

There is little need for comments inside small Verilog library modules. a one-line comment preceding the module is all that is typically needed. One block of header comments (see section 6.2.1.4) at the beginning.

6.5 Editing Practices

Some readers may regard editing practices as mundane and discussed too often in the different books available on Verilog, hardware description languages, and programming languages in general.

The authors feel that consistent editing practices, particularly in larger design projects, can be as critical to RTL verification success as any other topic in this book. The importance of consistent, good-looking Verilog to verification success makes it worthwhile to repeat the basic editing practices in this book. To those who need to know the *why* behind an editing practice, we supply some reasoning behind them.

6.5.1 Indentation

Designers writing Verilog source code who follow source line indentation policies help the engineer reading the Verilog code to see the relationship of control statements.

Indentation Principle

A design project must define a uniform indentation policy.

Keating and Bricaud [1999] suggest using 4 spaces for indentation and not tabs. They point out that treatment of tabs is not uniform across various source viewers that different engineers may need to use to read a given Verilog source. On top of the eternal split between the choice of *vi* and *emacs* by different engineers, some new verification tools supply their own source viewer and editor that is neither *vi* or *emacs*.

On the other hand, tabs may be no problem to the engineers reading Verilog with their various source viewing and editing tools. If so, use of tabs in place of multiple spaces for indentation can save file space. In one experiment, we found that tabs in place of spaces for indentation reduced the file space for the 60,000 lines of RTL representing a 300K -gate ASIC from 16 Mbytes to 11 Mbytes. Measuring simulation compilation time for these two files showed no difference between the time required for parsing the spaces and the tabs for indentation.

In spite of the file space argument, the authors believe that designers must adopt the habits that support reuse, and use spaces instead of tabs even for modules in projects that have no possible use outside the project domain.

6.5.2 Comments

Before going into detail about specific classes of comments, let us first state that comments must be supplementary to the actual Verilog code, and must not overwhelm it. A reasonable limit on the amount of text devoted to comments is around 30% or less.

Place comments to the right of Verilog statement, or in a block preceding the lines to which they apply. This practice allows engineers reading the Verilog to focus on the code itself, or on the comments for supplementary understanding.

Comment blocks must describe higher-level aspects of the Verilog that follows, and not merely repeat in prose exactly what the Verilog statements express.

Small library modules need few, if any comments, since the Verilog should be short and self-explanatory. A single header block at the beginning of the multi-module library file and a short descriptive header preceding each module should suffice.

6.5.2.1 Header

Comments identifying the design content of a design file must be placed at the beginning of each design file. The comments must include

- Copyright notice. The beginning of protection for the author and employer.
- Authors name(s). Tells engineers looking at the file who can answer questions regarding the file. If the author(s) are no longer available, engineers can upwardly adjust the cost estimate of changes and verification accordingly
- Date written. Knowing the date that the file was originally written helps engineers more accurately estimate the cost changes and verification. In general, the older the design, the less likely it will be to find anyone with fresh knowledge of the design trade-offs in the original design.
- A short description. Provides the key information that will help engineers get started in their understanding of the design file contents.
- Revision history. Following a revision identifier, the author, date, and short description must be included with each revision, just as with the header for the file.

Automated revision control systems provide help with most of the above the bookkeeping details. The author must provide thoughtful entries beyond automated boiler plate. One key item in the revision description is crediting the tool or process that prompted the revision. This credit helps later in the project when designers want to know what tools and processes are providing the most verification payoff.

6.5.2.2 Declarations

Comments on the right-hand end of the line that expand on the abbreviation and function of a declared port or signal are effective. Here is an example that expands on the meaning of port name declaration.

tabd_rreq, *// table data read request*

6.5.2.3 end Identification

Verilog designs that contain *end identification* comments are very helpful whenever engineers read someone else's Verilog that contains large blocks of code.

When viewing Verilog source with editors and source browsers, engineers often find themselves looking at the bottom of a block of code and having to page up in order to find the top of the block to see what block they are in. A comment at the bottom of the block that identifies the block makes this aspect of Verilog analysis work easier.

[Example 6-16] illustrates end identification comments for function, procedural block, and module.

Example 6-16

module respsend **(**

...

function [6:0] start_pointer;

...

endfunction *// start_pointer*

...

always @(r_pop_cnt **...**
   begin

   ...

   end *// always @(r_pop_cnt*

...

endmodule *// respsend*

6.5.2.4 Meta-Comments

The semantics of the Verilog language primarily relate to an event-driven simulator. With a few extra pieces of information, other EDA applications can use Verilog as their primary input.

Meta-comments[1] are ordinary //-prefixed or /* ...*/-bracketed comments that carry specialized information to Verilog-based applications. The specialized information is generally beyond that which can be carried within the semantics of the Verilog language.

The specialized information for cycle-base simulation, synthesis and code coverage applications has generally been simple off-on switches, or mode settings directing interpretation.

The following four examples are off-on switching comments that tell vendor code coverage or synthesis tools to exclude the code bracketed between these comments from coverage or synthesis tools.

// <vendor> coverage_off
// <vendor> coverage_on
// <vendor> synthesis_off
// <vendor> synthesis_on

---

1. Language design experts regard properties or attributes built into a language as superior to comments for communicating application-specific information to an extended set of applications. Version 2.0 of the [OVI LRM 1993] included Verilog extensions for attributes, but in the standardization process that resulted in [IEEE Std 1364 1996], the committee and the balloting decided to take properties out of the language definition.

The following example shows code exclusion directed by IEEE vendor-independent comments.

```
// rtl_synthesis off
  <test and diagnostic statements in Verilog>
// rtl_synthesis on
```

Here we have a comment that tells a simulation compiler to model the specified including the Z state even when compiling the rest of the design variables for cycle-based two-state simulation.

```
wire /*4value*/[31:0] tri_data;
```

Some formal model checking vendors have added proprietary property checking languages as comments within Verilog design files.

In general, it is a good idea to use standard, non-proprietary and vendor-independent meta-comments or other methods to send extra information to applications. Use of vendor-neutral source for a design allows a project maximum flexibility in its tool choices, and may facilitate re-use on future projects.

Here are some methods by which a project can keep their design source vendor neutral.

1. Use a preprocessor to translate from a vendor-neutral meta-comments to vendor-specific meta-comments.

2. Use IEEE standard meta-comments and keep asking the application vendor about the status of their support for IEEE standards.

3. Use Verilog **'ifdef** - **'else** -**'endif** where possible to serve the purpose of application off-on bracketing comments.

Meta-comment Principle

Avoid using vendor-specific meta-comments.

6.5.2.5 Embedded Comments

These are the hardest to write, but are potentially the most useful comments in a design file. They must not just repeat the Verilog code in prose form, but rather provide higher-level or alternative view of the Verilog functionality. [Example 6-17] illustrates a comment that expresses the functional intent of some Verilog source and in no way repeats the Verilog.

Example 6-17

```
// Determine if we can bypass. Unfortunately, need to wait for the
// A cycle in the table since the MC needs A cycle bits coincident
// with the win. Only local requests will be bypassed.
assign c_bp_hdr_val = (r_hdr_valids_1 == 11'b0) &&
                      (r_hdr_valids == 11'h400);
...
```

6.5.3 Line Length

In parallel with the move to larger screens on EDA workstations, increasing numbers of EDA tools require a multitude of windows be open at the same time. In many cases, these tools include a window for viewing and perhaps editing the Verilog source code.

Even though the 80-character limit on screens of the early 1980's are long past, an 80-character limit on line length is still a good idea. This limit allows entire lines to be seen within a smaller source window on a larger multi-window screen.

6.6 Summary

In this chapter, we specified ideas on style, and the reasoning behind them. The style ideas began with design content. The design content included asynchronous logic, combinational feedback, and case statements. The case statement section presented the arguments favoring the fully-specified case statement style to facilitate verification.

We then presented organization and naming conventions for the various elements of our verifiable RTL style, again with the reasoning in support of the style. An important factor in the naming of modules as well as user tasks and vendor library functions is support of simulation performance profiling, as well as avoiding clashes in their global name spaces during system integration.

We concluded with discussion of editing practices and their importance with respect to verification processes.

7
The
Bad Stuff

In previous chapters, we have tried to show good ways to write and use RTL Verilog to support verification processes. In this chapter, we look at specific examples of what projects, designers, and EDA verification tool developers have done that obstruct a productive verification process flow.

By explicitly pointing out the bad stuff, this chapter may be helpful to some readers of the preceding chapters who want to see what we are explicitly ruling out to achieve verifiable RTL design. Other readers may skip directly to this chapter with the intriguing title, and then read the preceding chapters that tell them what to do instead of the bad stuff.

Some of the bad stuff cited in this chapter is not all that bad, but is near the borderline between what we would consider verifiable RTL and not so verifiable RTL. Some are a matter of degree that might not hurt verification, like using a couple extra carefully-selected keywords from the unsupported set, or making a few bit-references to a bus.

Others are purely a matter of arbitrary choice. Where there are three different ways to do the same thing in RTL Verilog, we pick one, and relegate the other two to this bad stuff chapter. Leaving the choice up to each designer on a project team may seem to provide an initial gain in designer productivity. This productivity gain, however, is overwhelmed by the increased costs of reading and supporting a wide range of constructs by verification engineers and EDA tools.

Examples of the very bad stuff include:

- expressing flip-flops as in-line code instead of objects
- using the X-state in RTL Verilog
- killing RTL simulation performance by frequent and bit-level visits
- using constructs that cause simulation differences between the RTL and the synthesized gate model
- writing Verilog that has logic timing problems, where a resultant erroneous state is dependent upon a particular sequence of events
- vendor EDA tools that break their customer's verification process flow
- design teams that do not define and follow a verification-oriented process
- drawing keywords and statement types from the entire Verilog language in RTL design
- user-defined primitives, and especially sequential user-defined primitives

7.1 In-Line Storage Element Specification

[Example 7-1] (a) illustrates a familiar RTL coding style that specifies flip-flops in-line. It is bad because it locks in on a flip-flop description style, which hinders adaptation to design and verification tools.

Example 7-1

a) Bad: Flip-flops in-lined in module	b) Good: Flip-flops instantiated in module

```
always @(posedge ck250)
begin
    r_rcs <= rst_ ? c_rcs : 0;        dff_r reg_rcs (r_rcs, ck250, rst_, c_rcs);
    r_del <= c_del;                   dff_r5 reg_del (r_del, ck250, c_del);
    r_avail <= c_avail;               dff reg_avail (r_avail, ck250, c_avail);
    r_n1 <= rst_ ? c_n1 : 0;          dff_r5 reg_n1 (r_n1, ck250, rst_ , c_n1);
    r_n2 <= rst_ ? c_n2 : 0;          dff_r5 reg_n2 (r_n2, ck250, rst_ , c_n2);
    r_n3 <= rst_ ? c_n3 : 0;          dff_r5 reg_n3 (r_n3, ck250, rst_ , c_n3);
    r_n4 <= rst_ ? c_n4 : 0;          dff_r5 reg_n4 (r_n4, ck250, rst_ , c_n4);
    r_n5 <= rst_ ? c_n5 : 0;          dff_r5 reg_n5 (r_n5, ck250, rst_ , c_n5);
    r_n6 <= rst_ ? c_n6 : 0;          dff_r5 reg_n6 (r_n6, ck250, rst_ , c_n6);
end
```

[Example 7-1] (b) is good because it isolates tool-specific details about flip-flop modeling within tool-specific libraries. This methodology facilitates simultaneously optimizing the performance of simulation, equivalence-checking, model-checking and physical design within a project's design flow. See chapter 3 for a complete explanation.

7.2 RTL X State

Two-state in this book refers to eliminating the X, and using only 0, 1 and Z states. Although tri-state buses have an important place in modern system design and simulation, the bulk of the logic and nodes are only two-state, not tri-state.

Our initial purpose in eliminating the fourth X-state was simulation performance. We are not alone in eliminating the X. In recent years, new vendor simulator releases provide the option of simulating without an X-state in order to achieve greater simulation performance.

However, we believe that using the X-state in RTL simulation is a bad idea, even without the performance penalty that it causes. RTL simulation using the X-state can be both excessively pessimistic and optimistic, and attempts at overcoming these shortcomings are impractical.

7.2.1 RTL X-STATE PROBLEMS

7.2.1.1 RTL X-State Pessimism

Arithmetic operations are one example of gross pessimism in X-state RTL simulation. Consider the [Example 7-2].

Example 7-2

```
reg [15:0] a,b,c;
    ...
    begin
      b = 16'b0000000000000000;
      c = 16'b000000000000X000;
      a = b + c;
      $display(" a = %b",a);
    end
```

The result for "a" in a four-state Verilog simulator will be

"a = XXXXXXXXXXXXXXXX".

In RTL simulation of arithmetic operations, fast simulators map these operations into host computer instructions. These fast simulators detect any X-bits in the input operands by checking an extra "flag word" for each input operand. Bits that are "1" in the "flag word" mark bit positions that are X in the input operand. So if the flag word is non-zero for either input operand, the simulator skips the addition instruction, and assigns all X's to the result. Note that the overhead added by the check for X-bits in an input operand is a single-instruction step, and therefore closely matches the performance of a single host-machine arithmetic instruction.

At the cost of reduced simulation performance, a Verilog gate-level simulation can more accurately handle this addition, resulting in "a = 000000000000X000". The gate level simulator can propagate the X more accurately because it pays the performance cost of visiting each bit in each operand, and generates a result bit-by-bit.

[Example 7-3] illustrates pessimism in a **case** statement.Consider the situation where the control signal "d" is "0X." Interpreting the "X" as a possible "0" or "1," only the first two case branches should be reachable. So, less pessimistically, only the left bit of "e" is ambiguous, and the result should be "e = X1." However, a four-state Verilog simulator will give "e = XX" when control signal "d" is "0X."

Example 7-3

```
reg [1:0] d,e;
    ...
    begin
    d = 2'b0X;
    case (d)
      2'b00 : e = 2'b01;
      2'b01 : e = 2'b11;
      2'b10 : e = 2'b10;
      2'b11 : e = 2'b00;
      default : e = 2'bXX;
    endcase
    display(" e = %b",e);
    end
```

7.2.1.2 RTL X-State Optimism

More insidious is the way that RTL simulation of **case** statements and **if-else** statements with an X-state can lead to optimistic results, and thereby hide real start-up problems in a design.

Given an XX as the start-up state for d, the **case** statement in [Example 7-4] will take the **default** branch. That only test one of the four possible branches the start-up condition could actually take, if we consider the four possible two-state interpretations of the XX bits.

Example 7-4

```
reg [1:0] d,e;
    ...
begin
  case (d)
    2'b00 : e = 2'b01;
    2'b01 : e = 2'b11;
    2'b10 : e = 2'b10;
    default : e = 2'b00;
  endcase
  $display(" e = %b",e);
end
```

7.2.1.3 Impractical

As a thought exercise, it is possible to envisage an RTL style that would intercept and process X-states more accurately, moderating both the pessimism and the optimism.

[Example 7-5] (a) shows an **if-else** statement that accurately intercepts and propagates an X-state. [Example 7-5] (b) presents a **case** statement that is similarly modified to intercept X-states and propagate their affect on the result more accurately.

Example 7-5

a) X intercept in if-else

```
if (f = = = 1'b0)
  g = 2'b00;
else
if (f = = = 1'bX)
  g = 2'b0X;
else
  g = 2'b01;
```

b) X intercept in case

```
reg [1:0] d,e;
    ...
begin
  case (d)
    2'b00  : e = 2'b01;
    2'b0X  : e = 2'bX1;
    2'b01  : e = 2'b11;
    2'bX0  : e = 2'bXX;
    2'bXX  : e = 2'bXX;
    2'bX1  : e = 2'bXX;
    2'b10  : e = 2'b10;
    2'b1X  : e = 2'bX0;
    2'b11  : e = 2'b00;
  endcase
end
```

Another way around the pessimism/optimism problems with the **case** and **if-else** statements is to express the state transitions in boolean form. [Example 7-6] shows how the state transitions in [Example 7-4] can be expressed in a boolean form that propagates X's with only the mild pessimism familiar to users of X-state in gate-level simulators.

Example 7-6

```
reg [1:0] d,e;
    ...
begin
    e = { ( ^ d), ~d[1]};
end
```

These examples illustrate how RTL usage that attempts to intercept X's everywhere is a not a good idea. Here are some reasons for not intercepting X's.

- Simulation performance. For **case** and **if-else** statements, all the extra tests for X's add to the CPU processing that the simulator has to do.

- Labor content. Someone has to do the work of adding the extra X-test **case** and **if-else** statements, or reduce the branch statements to boolean form.

- Complexification. A good feature of RTL design is that it can present a designer's intent more clearly than boolean-level design, and intercepting X's detracts from the clarity.

- Completeness. There is no current method of guaranteeing that the designer's X interception and propagation is complete enough to avoid the pessimism and optimism.

- Synthesis. X interception makes the RTL a ternary logic design, which has to be thrown out when mapping the design to binary logic gates in synthesis.

We prohibit use of X-intercepting and X-assignments anywhere in our RTL logic design. This includes the X-intercepting default in fully specified case statements as shown in Thomas and Moorby [1998] and in [Example 7-7].

Example 7-7

```
    ...
case (select)
    2'b00 : mux = a;
    2'b01 : mux = b;
    2'b10 : mux = c;
    2'b11 : mux = d;
    default : mux = 'bX;
endcase
```

Our RTL design style requires that all **case/casex** statements be fully-specified, so assigning an X in a default is never needed for telling synthesis about don't-care situations.

Contemporary logic synthesis technology allows for greater optimization of generated gates for **case/casex** statements in which certain input control variable state values are impossible. For these **case/casex** statements, the designer does not care about what output states the gates generate for those control state values.

Given the importance that we assign to RTL-based verification, we feel that the extra gates saved by allowing synthesis to optimize don't-care logic are not worth:

- precluding the simulation of gate-based ATPG test vectors against the RTL chip models.
- the challenges it presents to fast RTL-to-gate boolean equivalence checking between the RTL and the gate level description [Foster 1998].
- the semantic mismatches between RTL and gate-level simulation.

7.3 Visits

Chapter 4 introduced the principle of minimizing the frequency and granularity of visits for best RTL logic simulation software performance. In this section, we review RTL styles that degrade simulation performance by their high visit frequency and fine visit granularity. Primary visit simulation performance offenders include:

- referencing bits instead of buses,
- configuration tests throughout the duration of a simulation, and
- loops.

7.3.1 Bit Visits

To achieve the best RTL simulation performance, designers writing Verilog code focus on the signal bus instead of the signal bit. In Chapter 6, we recommended parallel value operations instead of operations on individual bits. In that chapter, we used the example of a content-addressable memory coding. In [Example 7-8], we illustrate the Verilog coding for error-correcting encoding logic, using bit references (a), which simulate slow (bad), and parallel value operations (b), which simulate fast.

Example 7-8

a) Bit references   **b)** Parallel value operations

```
c_ecc_out_1 =c_in [10] ^ c_in[11]        c_ecc_out_1 =
          ^ c_in[12] ^ c_in[13]                ^ (c_in & 40'h003ffff893);
          ^ c_in[14] ^ c_in[15]
          ^ c_in[16] ^ c_in[17]
          ^ c_in[18] ^ c_in[19]
          ^ c_in[20] ^ c_in[21]
          ^ c_in[22] ^ c_in[23]
          ^ c_in[24] ^ c_in[25]
          ^ c_in[26] ^ c_in[27]
          ^ c_in[28] ^ c_in[32]
          ^ c_in[35] ^ c_in[38]
          ^ c_in[39];
```

Note that [Example 7-8] (b) is a more of *register* transfer operation. Its compactness makes the functional intent more clear and obvious to a reader, in addition to simulating faster.

7.3.2 Configuration Test Visits

A project can improve simulation performance by eliminating configuration test visits after simulation start up. Move configuration decisions to:

- compilation controlled by **'ifdef** -**'else** - **'endif.**
- text macro preprocessing (as described in Chapter 3), or
- instantiation of distinct library module types for each distinct functionality.

Consider the [Example 7-9] of a parameterized first-in-first-out (FIFO) queue model that designers instantiate in different flavors throughout a design. The instances differ in their width, depth and whether to encode the one-hot data input. With every different value written to the queue, the model calls the encoder function and returns the indata or the encoded version of indata. This call costs in simulation run time with every write to the queue.

Example 7-9

```
module fifo(
...
parameter WIDTH = 13;
parameter DEPTH = 32;
parameter ENCODE = 0;
...
function [31:0] encoder;
input  [WIDTH-1:0]   indata;
begin
  if (ENCODE != 0) begin
    < calculate encode value based on indata >
    end
  else
    encoder = indata;
end
```

The simpler and better way is to define two FIFO library types, one that encodes its data input, and another that doesn't. Just as with a parameterized FIFO module instance, the decision as to whether to use an encoding version or not is on the instantiation line.

```
fifo #(12, 64, 1) iqueue (...); // Bad, parameterized functionality
fifo_e #(12, 64) iqueue (...); // Good, functionality decided at compile time
```

The separate models for each functionality makes the models easier to understand, and more likely to simulate correctly as well as fast.

7.3.3 for Loops

In our experience, the only RTL need for a **for** loop is memory array models that have a *clear memory* functionality. Since the OOHD methodology (see chapter 3) encapsulates memory in library modules, we limit the **for** loop to the library designer, and do not make it available to the chip designer.

Widespread use of the **for** loop degrades simulation performance when designers misapply it. Sampling Verilog from projects that allowed the **for** loop in chip designs, we found that every non-memory **for** could be eliminated, and the *clear memory* **for** loop could be rewritten to achieve far better simulation performance.

[Example 7-10] presents an example of a **for** loop from a real design, and the simpler, faster, clearer way to write the same logic. In the bad example, notice how there is a count increment, a test for loop completion, and a visit to every bit. The good example eliminates the loop overhead, and allows the simulator to act on the bits in parallel, loading, inverting and storing the host machine word..

Example 7-10

a) Bad: Slow / Less Obvious

```
input ['N-1:0] a;
output ['N-1:0] b;
integer i;
reg ['N-1:0] b;
always @ (a) begin
  for (i=0; i<='N-1; i=i+1)
    b[i] = ~a[i];
end
```

b) Good: Fast / Simple and Clear

```
input ['N-1:0] a;
output ['N-1:0] b;
assign b = ~a;
```

It is often impossible for even the writers of the original Verilog to determine what would lead to their using a **for** loop like [Example 7-10] (a). We can guess that they had a "gate-instantiation" viewpoint instead of a RTL viewpoint at the time that they wrote the Verilog.

Compared to [Example 7-10], it may appear somewhat legitimate to use a **for** loop at interfaces between opposite bit-ordering conventions. [Example 7-11] shows a loop-based bus reversal and a concatenation-based bus reversal.

In good design practice, the need for bus bit-ordering reversals is very rare. It might be argued that because they are rare, their simulation performance effects would be small. Amdahl's Law [Amdahl 1967], however, warns that slow parts of a process will tend to dominate in the overall process performance. Their rarity also means that the productivity gain from using the **for** loop instead of concatenation would be very minor.

Example 7-11

a) Bad: Simulates slower

```
input [15:0] a;
output [0:15] b;
integer i;
reg [0:15] b;
always @ (a) begin
  for (i=0; i<=15; i=i+1)
    b[15 - i] = a[i];
end
```

b) Good: Simulates faster

```
input [15:0] a;
output [0:15] b;
assign b = {a[0], a[1], a[2], a[3],
  a[4], a[5], a[6], a[7],
  a[8], a[9], a[10], a[11],
  a[12], a[13], a[14], a[15] };
```

[Example 7-12] (a) shows a FIFO memory model example with poor simulation performance. Here are some of its performance problems.

- Putting the **for** loop outside the case results in repeatedly testing whether reset is on or off.

- Rewriting all of the unaddressed words takes simulation time and contributes nothing to the function's verification.

[Example 7-12] (b) shows the same FIFO memory model with improved simulation performance. It tests for reset only once, and only loops if the reset is true. It also eliminates the rewriting of the unaddressed memory words.

Example 7-12

a) Memory model with poor simulation performance

```
for(i = 0; i < fifo_depth; i = i+1)
   begin
     case({reset_L_ff, w_addr_ff == i})
           2'b00,
           2'b01: entry_ff[i] <= 0;
           2'b11: entry_ff[i] <= write_data;
           2'b10: entry_ff[i] <= entry_ff[i];
     endcase
   end
```

b) Memory model with improved simulation performance

```
if (reset_L_ff)
    for(i = 0; i < fifo_depth; i = i+1) entry_ff[i] <= 0;
    else
    entry_ff[w_addr_ff] <= write_data;
```

7.4 Simulation vs. Synthesis Differences

This section describes Verilog RTL coding styles that yield mismatches between RTL simulation and post-synthesis gate-level simulation. Mills and Cummings [1999] aptly contend "that any coding style that gives the HDL simulator information about the design that cannot be passed on to the synthesis tool is a bad coding style. Additionally, any synthesis switch that provides information to the synthesis tool that is not available to the simulator is bad." To prevent mismatches between RTL and post-synthesis simulation, both processes must possess equal understanding of the RTL design model. We restate this idea as the *Faithful Semantics Principle*.

Faithful Semantics Principle

A RTL coding style and set of tool directives must be selected that insures semantic consistency between simulation, synthesis and formal verification tools.

To avoid RTL and gate-level simulation differences, design projects can adopt the RTL Verilog style presented in this book. They must enforce the style by tailoring a lint tool rules set, and locking the linting step into their design process to check all RTL Verilog.

If a project does not enforce faithful semantics, RTL simulations lose their credibility, and much more gate-level simulation is required. Because equivalence checkers base their RTL semantics on synthesis RTL policies, they are generally no help in detecting RTL simulation and synthesized gate simulation differences.

The following RTL simulation and synthesized gate simulation differences draw from our own experience and Mills and Cummings' [1999] paper. Their paper tells story after story of bad silicon resulting from designers overlooking RTL simulation and synthesis differences. We are in complete agreement with their goal of avoiding these differences, and carry this one step further by not allowing the X-state in RTL Verilog simulation.

We divide the causes of differences into three categories:

- explicit differences,
- careless coding, and
- timing.

7.4.1 Explicit Differences

RTL-based Verilog simulation and synthesis tools allow designers to deliberately go awry in their RTL verification process, and create differences between the RTL and gate-level simulation behaviors.

7.4.1.1 Full and Parallel Case

The *full_* and *parallel_case* synthesis-directing comments provide more information to the synthesis tool than used by the RTL simulator. They too often result in gates that don't simulate the same as the RTL.

Full case. In Chapter 6, we presented the verifiable RTL design requirement of fully-specifying **case/casex** statements in the RTL, using the *full_case* [Example 6-2] (a). Let us look again at this example here in [Example 7-13] and consider what goes wrong in RTL simulation.

Example 7-13

```
module c (r_o, c_n);
    input [1:0] r_o;
    output [1:0] c_n;
    reg [1:0] c_n;
    always @(r_o)
      case (r_o)  // rtl_synthesis full_case
        2'b00 : c_n = 2'b01;
        2'b01 : c_n = 2'b10;
        2'b10 : c_n = 2'b00;
      endcase
endmodule
```

Given the synthesis-directing *full_case*, contemporary synthesis tools generate gates that assign 0 or 1 values to the two bits of c_n for the case when r_o has the value 2'b11. Synthesis optimizations choose the value for c_n in this case.

On the other hand, the RTL simulator treats c_n as a latch when r_o has the value 2'b11. The designer may contend that the 2'b11 for r_o is impossible in normal operation, but there are circumstances that the designer must consider when making that contention:

- states during the start-up sequence,
- scan state sequences, and
- the designer's contention may be wrong.

Designers can add an assertion for the impossible r_o in the 2'b11 state and get diagnostic messages to deal with the normal operation. However, the assertion for the 2'b11 state does not address problems with the start-up sequence and scan operation.

Parallel case. For **casex** statements that have overlapping case-item constants, the *parallel_case* synthesis directive produces gates that do not simulate the same as the RTL.

For **casex** statements with unique non-overlapping case-item constants, the simulation behavior is the same between the gates and the RTL. Whether the *parallel_case* synthesis directive is present or not, synthesis produces the same gates for this class of **casex** statements. So the *parallel_case* synthesis directive is superfluous.

[Example 7-14] shows a **casex** statement implementation of a priority encoder. It includes the added *parallel_case* that tells synthesis to produce faster logic based on the assumption that only one bit of c_hot is 1. In response to the *parallel_case*, the synthesized logic behaves like a multiplexer, selecting one of the values, and or'ing it with the non-selected paths. The gate simulation matches the RTL simulation only within the bounds of the assumption. In situations where more than one bit is 1, the gate-level version or's the assigned c_code values, while the RTL version still simulates as a priority encoder selecting only one assigned value for c_code.

Example 7-14

```
casex (c_hot) // RTL synthesis parallel_case
    8'b1???????: c_code = 3'b000;
    8'b?1??????: c_code = 3'b001;
    8'b??1?????: c_code = 3'b010;
    8'b???1????: c_code = 3'b011;
    8'b????1???: c_code = 3'b100;
    8'b?????1??: c_code = 3'b101;
    8'b??????1?: c_code = 3'b110;
    8'b???????1: c_code = 3'b111;
endcase
```

Just as with the *full_case* synthesis directive, use of the *parallel_case* synthesis directive too often is based on the same assumptions that turn out to be wrong.

Eliminating full and parallel case. Here are the style elements that eliminate the *full_case* and *parallel_case* synthesis directives, and thereby maintain alignment between RTL and synthesized gate simulation behavior.

- Fully-specified **case/casex** statements. For **case/casex** statements, this means enumerating all case-item constant values, with either explicit constant values, a **default** within the **case/casex** statement, or a default value assignment preceding the **case/casex** statement.

- Eliminating all overlaps from case-item constant values. In the [Example 7-14], replacing all of the '?'s to the left of the '1' with '0's eliminates the overlaps.

- Accepting the priority encoder in gates that synthesis generates, with its added timing delays and gate count. This makes the simulation behavior of the gates match that of the priority encoder in the RTL. In non-critical delay paths and areas where gate-count is not significant, a priority encoder in gates is perfectly acceptable.

- Explicitly specifying a multiplexer in the RTL. Implementing the RTL priority encoder as multiplexer (see [Example 6-4] in Chapter 6) makes the RTL simulation match the gate simulation, as well as minimizing the delay and gate count.

7.4.1.2 X Assignment

In addition to all of the pessimism, optimism and impracticality problems of RTL X-state simulation discussed earlier in section 7.2 of this chapter, we also remind the reader that it causes simulation differences between the RTL and the gate-level.

Although it is possible to craft RTL logic in terms of boolean expressions in place of **case/casex** and **if-else** statements to make the X-state propagate more accurately, such crafting is counterproductive. To be completely safe in their X propagation, designers have to rule out their use of **case/casex** and **if-else** constructs from their Verilog RTL design style.

Many designers believe that making 'X' assignments for unused states in RTL state machine design is a useful trick for debugging bogus state machines. Because they see this trick working for them on many of their state machines, it is a strongly-held belief.

However, hard-earned experience with bad silicon caused by the RTL X-state optimism on other projects, and repeated success with good silicon on our projects using two-state RTL simulation with random initialization has convinced us that any crafting of the X-state in the RTL is misguided.

It is better to eliminate thinking about the X in RTL Verilog, and focus the project's Verilog style towards the fastest cycle-based, two-state RTL simulation possible. Random initialization in the cycle-based simulator can bring out the start-up problems previously thought to be addressed by the X-state in standard Verilog RTL simulations.

Our method is to run 99% of our simulations at the RT-Level using cycle-based, two-state techniques with random initialization, and 1% of our simulations at the gate-level with X's. So far, this method has caught all of our start-up state problems before silicon. We usually detect and fix one last start-up state problem for each new chip design using X-state simulation at the gate-level before going to first silicon [Bening 1999b].

7.4.1.3 Other Forms of State Machine

To the other forms of state machines, we apply the *Verifiable Subset Principle* (see Chapter 3). Applying this principle to a design project using two-state RTL simulation, the **case** and the **casex** (for its wild card) with control variables are a simple and sufficient subset. This policy rules out:

- constant case test expressions,
- implicit state machines, and
- **casez**.

Each of these adds to the complexity of the Verilog, and have their own peculiar ways of compounding the complexities of simulation differences between the RTL and gate-level.

Constant case test expressions. Case statements with a constant (typically a '1') in their test expression combined with *parallel_case* as shown in [Example 7-15] produce RTL and gate-level simulation differences in the same way as the **case/casex** statement with a controlling signal as shown in [Example 7-14].

Example 7-15

```
casex (1'b1) // RTL synthesis parallel_case
    c_hot[7] : c_code = 3'b000;
    c_hot[6] : c_code = 3'b001;
    c_hot[5] : c_code = 3'b010;
    c_hot[4] : c_code = 3'b011;
    c_hot[3] : c_code = 3'b100;
    c_hot[2] : c_code = 3'b101;
    c_hot[1] : c_code = 3'b110;
    c_hot[0] : c_code = 3'b111;
endcase
```

Since constant case test expressions are another way to say the same thing, we follow the *Verifiable Subset Principle* (see Chapter 3) and rule it out.

Implicit state machines. [Example 7-16] illustrates Verilog code for an implicit state machine. Synthesis tools support implicit state machines. Implicit state machines eliminate the **case/casex** statement from the state machine and merge the state transitions into the single flow of control. While Arnold *et al.* [1998] described and advocated implicit state machine techniques in Verilog, they noted that designers need to take care to avoid simulation and synthesis differences.

Example 7-16.

```
always
  begin
    @(posedge ck);
        e1 <= 2'b00;
    @(posedge ck);
        e1 <= 2'b01;
    @(posedge ck);
        e1 <= 2'b11;
    @(posedge ck);
        e1 <= 2'b10;
  end
```

Implicit state machines do not fit into our verifiable RTL design style for two reasons:

- They introduce another set of complex rules regarding potential RTL simulation and synthesized gate simulation differences.
- They merge the designer's functional intent with the state machine state register storage.

casez. In its two-state semantics, the **casez** is exactly the same as the **casex**. They both provide the very useful wildcard '?' "don't care" option for case-item constants. Based on alphabetical order, we picked the **casex** to support, and avoid any issues regarding differences in RTL and gate-level simulation of the **casez**.

In verifiable RTL design style, we accommodate the Z-state by encapsulation and assertions as described in section 6.1.4 of chapter 6. The encapsulation methods for tri-state receivers shown there provide better verification than the "don't care" treatment of the Z-state in **casez** statements.

7.4.1.4 Initial blocks

Designers generally enclose initial blocks between translate off/on directives. This method explicitly gives more information the RTL simulation than to the synthesized gate-level simulation. The designers generally do this to temporarily bypass start-up sequence testing, and go straight to testing the post-reset functionality of the block, chip and systems.

Example 7-17

```
module dff (q, d, ck);
    output [7:0] q;
    input [7:0] d;
    input ck;
always @(posedge ck)
    q <= d;
// rtl_synthesis off
initial
    q = 8'h00;
// rtl_synthesis on
endmodule
```

[Example 7-17] places initialization code directly into the module. This method of bypassing initialization testing typically invalidates any later initialization testing with RTL simulation.

A better way to bypass or inclusion initialization testing is to make the decision conditional in the testbench, outside the chip design. This method localizes the control of whether a test runs with or without initialization testing.

The best way to control initialization is encapsulating it in an $Initial-State(q) user task called from within the **initial** procedural block, replacing the assignment to q. This localizes the decision-making as to whether to apply initialization within the user task, so that it does not have to be repeated in block, chip and system testbenches. This encapsulation is described in Chapter 3 as another one of the benefits resulting from the OOHD methodology.

7.4.2 Inadvertent Coding Errors

This section describes specific examples of inadvertent Verilog coding errors that can cause differences between the simulated behavior of the RTL Verilog and the gate-level Verilog. Designers with any experience in RTL Verilog quickly become familiar with all of these kinds of coding errors, which generally reinforces their locking linting into their design process as the first step.

7.4.2.1 Incomplete Sensitivity List

Incomplete sensitivity lists are the most well-known source of RTL simulation problems, RTL and gate-level simulation differences, and annoyance to logic designers. [Example 7-18] illustrates an incomplete sensitivity list, where the "*or z*" required for correct RTL event-driven simulation functionality is omitted.

Example 7-18

```
module b (p, w, x, y, z);
    input [7:0] w, x, y, z;
    output [7:0] p;
    wire [7:0] w, x, y, z;
    reg [7:0] p, r, s;
    always @(w or x or y) // or z omitted
      begin
        r = w | x;
        s = y | z;
        p = r & s;
      end
endmodule
```

7.4.2.2 Latch Inference in functions

Inadvertent latch inferences happen because of omitted default assignments in **if** and **case/casex** statements. Outside of functions, inadvertent latch inferences are indeed design errors, but they simulate the same in RTL and gate-level simulation models.

Within functions, inadvertent latch inferences due to omitted default assignments create RTL and gate-level simulation differences. The RTL function behaves as a latch in simulation, while the synthesize gates behave as combinational logic, with no state storage.

7.4.2.3 Incorrect Procedural Statement Ordering

Synthesized gates in the gate-level simulation behave as if a sequence of combinational logic statements is ranked ordered correctly, even where they are not. RTL simulations can behave as a latch, hanging on to previously assigned values for out-of-order assignments.

The procedural block in [Example 7-19] has an out-of-order assignment to p. If only y and z change within an evaluation cycle, the changes that they cause will not be seen on p until the next evaluation cycle. The simulation of the gates synthesized from this RTL propagates changes in y and z through to p within the same evaluation cycle.

Example 7-19

```
module b (p, w, x, y, z);
    input [7:0] w, x, y, z;
    output [7:0] p;
    wire [7:0] w, x, y, z;
    reg [7:0] p, r, s;
    always @(w or x or y or z)
    begin
        r = w | x;          // rank 1
        p = r & s;          // rank 2
        s = y | z;          // rank 1
    end
endmodule
```

In our experience, designers correctly sequence statements within procedural blocks as they initially write their Verilog over 99% of the times, but not 100%. Out of every 100,000 lines of Verilog, they may make one or two mistakes in their procedural statement sequencing. While RTL simulation may reveal these errors eventually, it is much more productive to detect them immediately after design entry through linting.

7.4.3 Timing

Verifiable RTL design requires that a design project encapsulate all Verilog containing timing or clock-generation. A project that allows RTL timing control decisions to be distributed throughout the team members will likely create difficulties in their verification process. In addition to RTL and gate-level simulation differences, other difficulties include:

- Haphazard use of delays in a design adds labor (or roadblocks) in progressing to cycle-based simulation and emulation.

- Logic races cause test differences (and failures) in moving a simulation from one vendor's simulator version to another version (or vendor).

- Verilog practices that use delays and introduce races in the RTL design complicate the timing verification of the RTL model. Commingling timing with the RTL function violates the *Orthogonal Verification Principle* (see Chapter 2).

7.4.3.1 Delays

Delay specification has no place in a designer's RTL Verilog. Because synthesis discards all delay values in the RTL, their use invariably results in confusion to the engineers reading the Verilog at a minimum, and differences between the simulation behavior of RTL and the synthesized gate-level logic. The following examples go from bad to worse practices.

Flip-flop assignment delays. The [Example 7-20] of delay usage in flip-flop model blocking assignments is fairly widespread practice.

Example 7-20

```
module dff_2 (q, ck ,rst , d);
    input clk ,rst;
    input [1:0] d;
    output [1:0] q;
    reg [1:0] q;
    always @(posedge ck)
        q <= #1 (rst == 1'b0) ? d : 2'b00;
endmodule
```

A feature of putting a delay in the nonblocking assignment is that it separates the controlling clock edge from the resultant q output change in a waveform viewer. Without the delay, the waveform display showing the clock and the data change appearing to happen at the same time is disconcerting to some designers.

A project must globally control the flip-flop assignment delay to be less than the clock period (to prevent long-path problems) and to be identical (for simulation efficiency). Data changes in q will be late if the delay #1 exceeds the clock period. If a project has many different delay values for flip-flop assignments, the simulator has to revisit all of the changed outputs at the different times when they change.

Flip-flop assignment delays may mask simulation event clock skew. This skew is not physical clock skew, but skew in the RTL simulation events.

Masking simulation event clock skew with unit delays in flip-flop non-blocking assignments is regarded as a feature in some design teams. However, we currently feel that skew in clock fanout paths reflects a poorly disciplined RTL design practice. It should be detected and cleaned out. Skew can sneak into the clock fanout paths of an RTL design when designers slip up and put non-blocking assignments or gate-level cells with "realistic delays" in their clock fanout paths.

In verifiable RTL design, the entire clock fanout must be either

- connections through ports,
- non-blocking assignments in procedural blocks, or
- assign statements.

Although we have used flip-flop assignment delays in past projects, we currently are against using them in our projects, because of the way that they mask inadvertent introduction of skew in the clock fanout.

It is important to note that if a project's leadership changes its mind about adding or removing the unit delay from all flip-flop assignments, the OOHD library-based technique localizes the change to the flip-flop library file.

Testbench delays. Engineers often write Verilog testbenches in a less disciplined manner than the way that they write Verilog for their chip designs. As shown in [Example 7-21] (a), they sometimes introduce delays to make the testbench insert control states or observe states just after a clock edge.

Example 7-21

a) Custom-timed inserted states   **b)** Common-timed inserted states

```
always @(posedge ck)              always @(posedge ck)
begin                             begin
   #0.005;                            `DELAY_I
   o_ad_valid  <= 2'bz;               o_ad_valid <= 2'bz;
   o_ad_validb  <= 2'bz;              o_ad_validb <= 2'bz;
   o_trans_id <= 6'bz;               o_trans_id <= 6'bz;
   o_master_id <= 3'bz;              o_master_id <= 3'bz;
end                               end
```

b) Clock-timed inserted states

```
always @(posedge ck_i)
begin
   o_ad_valid <= 2'bz;
   o_ad_validb <= 2'bz;
   o_trans_id <= 6'bz;
   o_master_id <= 3'bz;
end
```

Use of custom delay values to tune timing in test benches is not good for verification. It hinders application of cycle-based simulation by complicating the simulator's evaluation cycles. It also makes it impossible to synthesize the test bench into gates to include it in an emulation box along with gates for the chip design.

A better way is globally specifying the timing with a project-wide named constant delay for inserted states as shown in [Example 7-21] (b). Use of a named constant helps establish a project-wide time for inserting values, and allows for refinement of that time.

The best way is encapsulating the timing for observability and controllability within a special clock generator as shown in [Example 7-21] (c). This encapsulation supports both emulation synthesis and cycle-based simulation of the test bench along with the hardware design under test.

One disadvantage of using a special clock is the need to fan it out. Within the testbench module environment, the fanout probably is not a burden. But for test logging and assertion module instances within in the module hierarchy of the device under test, adding ports and connections to fanout the test clock timing is burdensome.

#0 delays. One form of timing control that is especially bad is insertion of #0 delays to fine-tune the event ordering for a particular simulator. These may work around a race for a particular version of a particular vendor's simulator, but too often get in the way migrating to the another version of a simulator.

7.4.3.2 Race Conditions

Logic races arise when engineers code their Verilog in a way that makes the resultant state dependent on the evaluation order of two procedure blocks triggered by the same event. The most frequent cases of logic races we have seen are in testbenches where the engineer used blocking assignments in two interrelated clock-triggered procedural blocks, as shown in [Example 7-22] (a).

The evaluation order in this example affects whether a simulation propagates changes from a to c in a single clock cycle or two clock cycles. If the first **always** block evaluates first, changes propagate from a to c in a single clock cycle. If the second **always** block evaluates first, changes propagate from a to c in two clock cycles.

The evaluation order in [Example 7-22] (a) cannot be guaranteed between two different vendor's simulators, or even successive version of Verilog simulators from the same vendor. Some people regard this as a bug in Verilog. In other viewpoints, it is a feature, since it allows enhancements to simulation performance to be unconstrained by a rigid evaluation order for simultaneous events.

Gates synthesized from [Example 7-22] (a) behave as though both assignments are non-blocking assignments as in (b), and take a second clock cycle to propagate changes from a to c.

Example 7-22

a) Blocking assignments with a race	**b)** Non-blocking assignments eliminate the race

```always @(posedge ck)```
```begin```
  ```b = a;```
```end```
```always @(posedge ck)```
```begin```
  ```c = b;```
```end```
	```always @(posedge ck)```
	```begin```
	  ```b <= a;```
	```end```
	```always @(posedge ck)```
	```begin```
	  ```c <= b;```
	```end```

c) Sequential/combinational blocks eliminate the race	**d)** Combined combinational block eliminates the race

```always @(posedge ck)```
```begin```
  ```b <= a;```
```end```
```always @(b)```
```begin```
  ```c = b;```
```end```
	```always @(a)```
	```begin```
	  ```b = a;```
	  ```c = b;```
	```end```

The (b), (c), and (d) in [Example 7-22] show ways of eliminating the race in (a). Each of them has a different behavior, but their state outcome is independent of a simulator's simultaneous event evaluation order.

(b) takes a second clock cycle to propagate changes from a to c.

(c) propagates changes from a to c in a single clock cycle.

(d) propagates changes from a to c in response to changes in a.

Engineers can use the newer race analysis features in waveform viewers, logic simulators, and static lint checkers to help detect and diagnose logic race conditions, then change their logic timing controls to more precisely specify the intended evaluation order.

7.5 EDA Tool Vendors

Design projects increasingly rely on EDA vendor tools for their success in design verification. In addition to contributing to the success in verification on design projects, the EDA vendor verification tools too often add difficulties to a project's verification process.

Good Vendor Principle

Verification tool vendors must support real user needs in a project's design environment, not the tool vendor's preferred environment.

The following three sections go into detail about difficulties we have encountered that could have been avoided if the vendors only knew ahead of time about the project's design environment. The difficulties include:

- library name clashes/profiling support,
- existing command-line/script "make" environments, and
- proprietary tool-directing comments.

To be fair, projects asking vendors to comply with the *Good Vendor Principle* should be complying with the *Disciplined User Principle* (see chapter 1) in their own work.

7.5.1 Tool Library Function Naming

Vendor simulation support library developers have generally not been completely aware of the scale of the system simulation models into which design projects link functions from multiple libraries. In our system simulation model executables, we have seen fifty or more function libraries linked with the compiled Verilog, totaling 30,000 to 120,000 functions.

As explained in Chapter 6, for avoiding integration name clashes and for profiling support within such a large name space, all functions must use a prefix common to all functions within each library.

Part of the quality testing and evaluation process at both the vendor and the user site should include review of the function entry point names, to ensure that they all have a common prefix. You can check for this in UNIX/LINUX environments by going to the library directory and entering:

nm libcv2c.a **| grep entry | grep -v static | more**

where libcv2c.a is a project-specific PLI library being examined. The list should contain names with a common prefix as shown below.

```
cv2c_report_percentages    |    2576|extern|entry    |$CODE$
cv2c_run_thread            |    2208|extern|entry    |$CODE$
cv2c_run_time_check        |    2280|extern|entry    |$CODE$
cv2c_run_tq                |    1208|extern|entry    |$CODE$
cv2c_stopwatch             |     480|extern|entry    |$CODE$
...
```

Simulation tool developers vary in the degree to which they apply good prefix-based naming practices in their library functions. Sampling some libraries at the time of this writing, we see that the library developers generally have used a prefix-based naming for their functions to some degree, as shown in [Table 7-1].

Table 7-1. Examples of Library function naming prefix usage.

Tool	Prefixed	No prefix
In-house library	159	18
Coverage	278	494
Wave Trace	479	83
Simulator	713	1881

We expect to see better use of prefixes on library functions in future releases of their simulation tool products.

7.5.2 Command Line Consistency

After Cadence opened the Verilog language in 1990, the first independent vendors began supplying simulators and support tools that closely followed the entire Cadence Verilog Reference Manual. This included the basic command line options, such as **+incdir, -f, -F** and **+define**. Users set up their scripts to run Verilog-based tools invoking these common command line options.

Since the arrival of these first Verilog-based tools, some vendors have departed from the original command line options to supply their own methods for invoking include directories, specifying file lists and defining compiler options. From our own experiences with vendor tool evaluations, we find that vendors depart from the original options for different reasons.

- Some tools have their origins in VHDL versions of a predecessor tool that the vendor extended for Verilog. Instead of a C, UNIX and script-based origin like Verilog, these tools reflect context semantics that are independent of a specific operating system.

- The original command line options are not in the IEEE 1364 [1995] standard. The tool developers who are not aware of the large investment users already have in the original options may think that adherence to these options is not important.

- It appears that some vendors are introducing frameworks so that their Verilog tools run in a consistent manner within their own domain.

Whatever the reasons, departing from support of the basic command line options is not a good idea.

- It adds to the set up time for evaluation and integration of new EDA tools. The scripts that support the standard option lists will not plug and play.

- The delay in setting up new tools may cause a user to run out of time for evaluation before fully realizing the advantages of the tools.

Vendors who listen carefully to their customers quickly get the message and generally add a method that supports these options, at least as an extension to their own manner of invoking their tool.

7.5.3 Vendor Specific Pragmas

Support for design tools beyond the Verilog language's original application to the Cadence Verilog XL™ simulator requires additional semantic information. Synthesis, coverage, and cycle-based simulation are examples of Verilog-based tools that include additional lines to direct them.

The common practice that has grown up across Verilog-based tools is the use of tool-directing comments, as shown below.

```
// rtl_synthesis off
// Diagnostic non-hardware Verilog code
// rtl_synthesis on
```

What is particularly disconcerting is the way that some Verilog tool vendors format their tool-directing comments to include their company or tool name.

```
// <vendor-name> coverage off
// Diagnostic non-hardware Verilog code
// <vendor-name> coverage on
```

For the user-oriented tool developers, the better way is to format their tool-directing comments in a form that is open, and a candidate for standardization. The IEEE Verilog RTL synthesis standardization group [IEEE 1364.1 1999] proposes tool-directing comments that do not specify the vendor or any proprietary tool name. This consensus-building should start at the original inception of the new tool with a standardization-oriented design.

Here is a general organization of a standardization-oriented tool-directing comment:

// rtl_<application-name> <application-keyword>

where *<application-name>* is a generic name for an application, such as **coverage** or **synthesis**.

7.6 Design Team Discipline

Poor design team discipline invariably drives up project costs, increases frustration with design tools, and results in project schedule delays. In the history of technology, certain teams of intelligent, creative engineers have been exemplary in following a high degree of design discipline. Other teams have included engineers who misdirected their creativity and upset the overall design and verification process flow.

Since each new project brings a new mix of engineers, design goals, and verification technologies, the project has to revisit the process by which they establish a design team discipline. The essential ingredient in the process is application of the *Disciplined User Principle* (see chapter 1)

Readers with prior RTL design project experience will recognize the varying degrees of designer discipline they have seen, and have their own horror stories. The following are some of our experiences with lapses in chip and module level design team discipline.

Chip level. A project had one chip design out of five that did not follow all of the verifiable design practices of the other four. Some of the deviations from the common design practices included:

- mixed upper and lower case in names,
- multiple modules per file, where the name of the file had no relationship to the names of any of the modules included,
- X-state assignments and tests, and
- incompletely specified **case/casex** statements.

Consequently, this chip could not use the full-chip fast RTL-to-gate equivalence check or the cycle-based simulation technologies. Other schedule delays came when other designers were hampered in their understanding of this chip as they tried to help with the design and verification.

After the frustrations in working on this chip, the project decided to spend the six labor-weeks to upgrade this chip design to the common verifiable RTL style used on the other four chips.

Module level. Out of the hundreds of modules that comprised each of the other four chips in the same project, there were two modules that used in-lined non-blocking procedural assignments instead of library module instances for flip-flops.

This meant that as the engineers added verification tool support changes to the flip-flop modules in the libraries, they had to remember to revisit the non-blocking procedural assignments outside the library in the chip modules. This was a minor hindrance in the project flow, so the project postponed corrective action for the next project.

Table 7-2. Verifiable RTL Unsupported Verilog Keywords

and	**highz1**	**rcmos**	**task**
buf	**ifnone**	**real**	**time**
bufif0	**integer**	**realtime**	**tran**
bufif1	**join**	**release**	**tranif0**
casez	**large**	**repeat**	**tranif1**
cmos	**macromodule**	**rnmos**	**triand**
deassign	**medium**	**rpmos**	**trior**
disable	**nand**	**rtran**	**trireg**
edge	**nmos**	**rtranif0**	**vectored**
endprimitive	**nor**	**rtranif1**	**wait**
endspecify	**not**	**scalared**	**wand**
endtable	**notif0**	**small**	**weak0**
endtask	**notif1**	**specify**	**weak1**
event	**pmos**	**specparam**	**while**
for	**primitive**	**strong0**	**wor**
force	**pull0**	**strong1**	**xnor**
forever	**pull1**	**supply0**	**xor**
fork	**pulldown**	**supply1**	
highz0	**pullup**	**table**	

7.7 Language Elements

7.7.1 Keywords

Some readers may not have read Chapter 3 and have skipped to this chapter directly. For them, [Table 7-2] repeats the list of Verilog keywords that we do not support for use in RTL chip designs.

As mentioned in chapter 3, most of these are for gate-level, not RTL design. Some that are not gate-level are more for test benches (**for, force, release**). The **for** also targets memory library elements.

7.7.2 Parameters

Although **defparm** and **parameter** are not in our bad keywords list, we discourage their use in RTL design. The main reasons that parameters run into troubles in a verification flow are:

- parameters often cause simulation run-time penalties, for configuration tests that could have been done at compile time, and
- the quality of parameter implementation varies between different vendor verification tools.

We favor use of '**define** and macro preprocessing (see chapter 3) for our design work. We use '**define** for specification of all constants. Where others might turn to parameters for their ability to specify constants per-instance, we use macro preprocessing.

Here are some specific examples of '**define** and macro preprocessing in place of parameters.

- Code inclusion controls

  Use of parameters for code inclusion control is generally very bad for simulation performance. For modules that have several functional variants, specify a separate library module type for each functional variant. Where there is a global inclusion control on all instances within a design, use '**ifdef**-'**else**-'**endif** controlled by a compiler option.

- Bit width, memory array sizes

  Where a design has the same function applied to different widths and memory array sizes for each instance, parameters may seem attractive. However, macro preprocessing can do the same per-instance width and size adjustments, and add the benefit of generating application-oriented libraries.

- **case, casex** statement state machine constants

  Since designers do not assign state machine constants per-instance, use of parameters for these constants is questionable. The '**define** provides the same constant definition by name capability for state machine constants.

7.7.3 User-Defined Primitives

Even though this book is about RTL design, and we rule out the gate-level keywords **primitive** and **endprimitive** from our RTL style, it is important to emphasize that, for verification processes, user-defined primitives (UDP's) are VERY bad stuff both at the RTL and gate-level.

Model checking and equivalence checking products generally support combinational UDP's in later releases. For sequential UDP's, verification product support remains questionable. Deriving the Boolean functionality from sequential UDP's is a far more difficult process than deriving Boolean functionality from combinational UDP's.

Successful RTL verification counts on the RTL design being the equivalent of the gate-level design. Sequential UDP's that impede RTL-to-gate-level equivalence checking wreck the whole RTL-based verification process, described in Chapters 2 through 5.

Compared with their competitors' projects that insist on UDP's in the Verilog, projects that completely eliminate UDP's from their Verilog are able to apply releases of advanced verification tools earlier and with more successful results. In our project work, we eliminate the vestigial UDP's by applying the library-based object-oriented methods described in Chapter 3.

Elimination of UDP's is a key application of the *Disciplined User Principle* presented in Chapter 1, by which projects can avoid problems with Verilog tools, especially formal tools.

7.8 Summary

Compared with most of the current publicly available books and papers on RTL design, the two most revolutionary ideas in this chapter are classifying:

- in-line flip-flop declarations, and
- the RTL X-state

as bad stuff. Through Barnes and Warren [1999], anonymous referee comments, and informal communication channels, we are aware of other design shops outside of our own company who have adopted encapsulated grouping of storage elements as well as thrown out the X from their RTL design.

On the other topics, there seems to be general agreement that the following are bad stuff:

- RTL and gate-level simulation differences,
- RTL styles that hamper simulation performance,
- poor design team discipline,
- EDA tool vendors who have no understanding of the users' environment,
- lack of RTL language element policies, and
- use of UDP's.

On one hand, designers are looking for specific examples of what is bad, and why it is bad. On the other hand, some readers in general agreement with many of the bad stuff ideas may disagree with details, or may have much more to add to some or all areas. That is as it should be. The authors have changed their minds about details during the writing of this book, and will continue to do so in the future. That is called *progress*.

8

Verifiable RTL Tutorial

This chapter presents a verification/user-oriented tutorial on behavioral register transfer level Verilog. To support good verification practices, it presents Verilog as a strongly-typed language with a keyword set limited to RTL design.

The focus on RTL verification distinguishes this tutorial from the many sources of Verilog training available in textbooks and courses. These other sources generally describe the Verilog language as specified in the IEEE 1364 [1995] standard, including delay, switch-level, gate-primitive, and X-state modeling.

The Verilog language specification directs Verilog tool implementors all right, but Verilog's poor type-checking, extensive levels of abstraction, and large keyword set makes the language, as specified, unsuitable for RTL verification. Use of Verilog in the RTL-specific style described in this chapter counts on a strong rule checking by a linting tool.

It is the authors' belief that engineers can be far more successful in completing their design by copying and modifying examples. These examples must meet the requirements of an entire design flow methodology, and emphasize verification.

This tutorial starts with a minimal design showing two statements, and then proceeds by adding examples of additional Verilog language statements. Statements consist of combinations of the language elements listed below.

- Keywords - shown in **arial boldface**, these are reserved by the Verilog language.
- Names - Begin with a-z, A-Z, and may contain these letters, numbers 0-9 or underbar _.
- Unsized decimal integers - Specify bit ranges, memory sizes. Examples: r_bit_range[7:0], reg [15:0] m_emory[0:511].
- Sized integers - use to represent bits. May be binary, octal, hex or decimal. Examples: 8'b0011_1001, 2'o2, 13'h0f, 9'd255. Maximum width: 256 bits. Bits - may have binary 0 or 1 values. Use "z" values only in binary sized integers to directly drive tri-state output or inout ports.

8.1 Module

8.1.1 Specification

The following [Example 8-1] illustrates the beginnings of a design in terms of the smallest possible Verilog module.

Example 8-1.

module b ;
endmodule

The Verilog language does allow you to make this file a little smaller by placing the above two lines on a single line, but that would violate a policy that we are starting with here and carrying through the rest of this book: place no more than one statement per line.

The user-assigned module name "b" in this example is an adequate name for personal desk-drawer experiments with the Verilog language, but for a real design project, you would use a more descriptive name or acronym that follows a project-wide policy.

[Figure 8-1] illustrates the module type "b" schematically as a square, with the type name designator at the bottom of the square.

Figure 8-1. Module "b" portrayed as a square.

The **module** and **endmodule** are Verilog reserved keywords. You may not use Verilog keywords as a name for a module, or any other name that you assign within your Verilog text. For example,

Example 8-2.

module module;
endmodule

is not legal.

8.1.2 Comments

There are two formats for entering comments in the Verilog language.

Example 8-3

// to end-of-line
/ to */*

For the most part, comments relay supplementary information from the design engineer that originally enters the Verilog text to other engineers on the project. Design projects generally have a standard template for comments that precede each module. Other comments specific to the design interspersed with the rest of the Verilog text should be written to supplement the design intent. They should supplement good layout and naming conventions, and not obscure the Verilog text itself.

Some comments relay information from the engineer to specialized tools that use the Verilog text. Standard Verilog language semantics define meaning in terms of an event-driven 4-state simulator. Use of Verilog in other types of tools (e.g., cycle-based simulators, synthesis, model checking) require more information than can be carried in standard Verilog reserved words and statements.

Throughout this tutorial, engineer-to-engineer Verilog comments are shown in italic text, while engineer-to-tool-directive comments are shown in bold italic text.

Here are some examples of the various forms of comments.

Example 8-4

```
/* The following module is a simulation
       diagnostic aid, not real hardware */
// rtl_synthesis off
module a ;
// The internals of this module will be supplied by the formal DV project
endmodule
// rtl_synthesis on
```

In formal terms, Verilog standards documents refer to a comment that communicates information from a user to a tool as meta-comment. The information within the comment is a pragma.

8.1.3 Instantiation

Modules may be instantiated within modules to form a hierarchy of modules. [Figure 8-2] presents the hierarchical diagram showing the relationship of module "b" with its submodule instances c1, c2, and d0 within b.

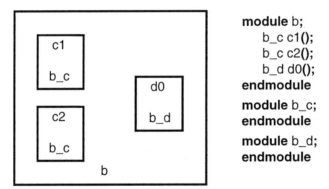

Figure 8-2. Module instances within a hierarchy

This example illustrates instantiations of module types b_c and b_d within module type b. The two lines starting with module type name b_c specify instance names c1 and c2 as the instance names for the two instances.

Also within module b, "d0" is the instance name for module type b_d.

8.1.4 Interconnection

Now let us look at how we can express the interconnection of submodules within a module in Verilog. To illustrate module interconnection in Verilog, we now introduce three more Verilog keywords: **input**, **output**, and **wire**.

Implicit Interconnection. Consider the interconnected submodules in [Figure 8-3]. This example shows the implicit style of expressing interconnection in Verilog. The input ports in each submodule are "i" and "j", while the output ports are called "o." For the module "b," the input ports are "w," "x," "y," and "z," and the output port is "p." The signal wires connecting c1 and c2 to d0 are "r" and "s," respectively.

The port order o, i, j defined for module type b_c in line 5 of the Verilog defines the connection of these ports to the wires in the instances b_c within module b.

- **wire** r connects to the ouput port "o" .
- **wire** w connects to input i.
- **wire** x connects to input j.

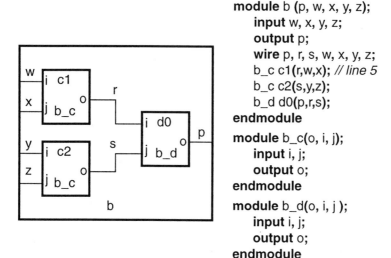

```
module b (p, w, x, y, z);
    input w, x, y, z;
    output p;
    wire p, r, s, w, x, y, z;
    b_c c1(r,w,x); // line 5
    b_c c2(s,y,z);
    b_d d0(p,r,s);
endmodule
module b_c(o, i, j);
    input i, j;
    output o;
endmodule
module b_d(o, i, j );
    input i, j;
    output o;
endmodule
```

Figure 8-3. Implicitly interconnected module instances

Explicit interconnection. The Verilog language also supports explicit identification of a submodule ports and the **wire**'s to which they connect. By explicitly naming each submodule port and the wire to which you want it connected to, the declaration order of the ports on the submodule is immaterial.

[Figure 8-4] illustrates the same interconnected modules shown in [Figure 8-3], but with explicit connections. You can see the contrast between implicitly specified connections and the explicitly specified connections.

Implicit: b_c c1(r,w,x); // *line 5*

Explicit: b_c c1(.i (w), .j (x), .o (r)); // *line 5*

In terms of the connections that they specify, these lines are exactly equivalent.

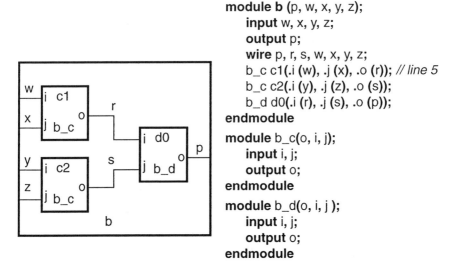

```
module b (p, w, x, y, z);
    input w, x, y, z;
    output p;
    wire p, r, s, w, x, y, z;
    b_c c1(.i (w), .j (x), .o (r)); // line 5
    b_c c2(.i (y), .j (z), .o (s));
    b_d d0(.i (r), .j (s), .o (p));
endmodule
module b_c(o, i, j);
    input i, j;
    output o;
endmodule
module b_d(o, i, j );
    input i, j;
    output o;
endmodule
```

Figure **8-4**. Explicitly interconnected module instances

Implicit vs. Explicit. The general rules for proper use of implicit and explicit connections of module instances are:

- Use explicit connection specification

  - on instantiations of modules that have:

    – complexity

    – multiple outputs

        – large number of inputs or outputs (more than five or so)

        – low usage

        – functionality for a specific part of a design in the project.

- • Wherever you (or anyone else) has any doubt as to whether to use explicit or implicit connections.

- Use implicit connection specification

  - • on instantiations of modules:

    – that have simple behavior.

    – that are single-output, with the output on the left end of the input-output list.

    – that have high usage.

    – that have functionality shared across the entire project.

    – that are from a library.

  - • only after the entire project is in prior and complete agreement about using implicit connections for the module.

8.2 Adding Behavior

Up until now, we have only looked at the Verilog language for describing structure. The keywords **input** and **output** relate to behavior, but not enough that we can simulate.

To illustrate behavior, we introduce the Verilog **assign** keyword, and our first use of a logic expression.

Example 8-5

```
module b_c(o, i, j);          module b_d(o, i, j);
   input i, j;                   input i, j;
   output o;                     output o;
   wire o;                       wire o;
      assign o = i | j;             assign o = i & j;
endmodule                     endmodule
```

The **&** and **|** operators in the above expressions perform the Boolean *and* and *or* functions on the 1 or 0 values carried by i and j. Using logic symbols for 2-*and* and 1-*or* in our hierarchy of modules, the overall behavior is shown in [Figure 8-5].

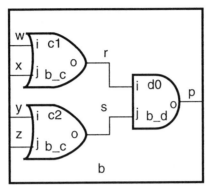

Figure 8-5. Schematic representing behavior.

8.3 Multi-bit Interconnect and Behavior

In the preceding sections, we have looked at specifying single-bit interconnect and behavior in Verilog.

Since the word register in RTL implies something multi-bit, let us next look at how to extend our preceding examples to a multi-bit functionality. In the following example, the declarations input, output and wire use the notation [7:0] to specify that the signals are eight bits wide.

Example 8-6

```
module b (p, w, x, y, z);
    input [7:0] w, x, y, z;
    output [7:0] p;
    wire [7:0] p, r, s, w, x, y, z;
    b_c c1(r, w, x);
    b_c c2(s, y, z);
    b_d d0(p, r, s);
endmodule
```

```
module b_c(o, i, j);
    input [7:0] i, j;
    output [7:0] o;
    wire [7:0] o;
        assign o = i l j;
endmodule

module b_d(o, i, j);
    input [7:0] i, j;
    output [7:0] o;
    wire [7:0] o;
        assign o = i & j;
endmodule
```

The assignment target o and the operands i and j in the above **assign** statements refer to all eight bits that comprise these signals. The follow examples show the equivalent bit-wise behavior represented by the *or* **assign** statement in module b_c.

Example 8-7

a) Expressed as a bit range

b) Expressed as individual bit-wise *or* operations:

assign o [7:0] = **i** [7:0] **l j** [7:0];

```
assign o [7] = i [7] l j [7];
assign o [6] = i [6] l j [6];
assign o [5] = i [5] l j [5];
assign o [4] = i [4] l j [4];
assign o [3] = i [3] l j [3];
assign o [2] = i [2] l j [2];
assign o [1] = i [1] l j [1];
assign o [0] = i [0] l j [0];
```

8.4 Expressions

Expressions in RTL Verilog consist of signals, constants, operators and parenthesis.

Parenthesis force the order of evaluation of an expressions, starting with the innermost parenthesis, and continuing outward. For example, the *and* and *or* behavior of the preceding hierarchy of modules examples could be (and more likely would be in RTL for real designs) written as a single **assign** statement within the module b.

Example 8-8

```
module b (p, w, x, y, z);
    input [7:0] w, x, y, z;
    output [7:0] p;
    wire [7:0] p, r, s, w, x, y, z;
        assign p = (w l x) & (y l z);
endmodule
```

8.4.1 Operators

In the following sections, we divide the 28 Verilog RTL operators into three groups:

- Binary - operate on two operands
- Unary - operate on a single operands

- Miscellaneous - multiple operators function as a set on two or more operands.

In the following three sections describing operators, **a** and **b** represent signals, constants, or the result calculated from a subexpression.

8.4.1.1 Binary operators

Arithmetic. Addition and subtraction is 2's complement. The operands **a** and
b must be the same width, and the result is the same width. The modulo **b**
operand must be a power of 2. The modulo result width is the same as **a**.

Operator	Context	Function
+	a + b	*addition*
-	a - b	*subtraction*
%	a % b	*modulo*

Bit-wise. The operands **a** and **b** must be the same width, and the result is the
same width.

Operator	Context	Function
&	a & b	*and*
|	a | b	*or*
^	a ^ b	*exclusive or*
~^	a ~^ b	*exclusive nor*

Logical. The operands **a** and **b** must be one bit wide, and the result is one bit
wide.

Operator	Context	Function
&&	a && b	*and*
||	a || b	*or*

Relational. The operands **a** and **b** must be the same width, and the result is
one bit wide.

Operator	Context	Function
==	a == b	*equality*
!=	a != b	*inequality*
>	a > b	*greater than*
<	a < b	*less than*
>=	a >= b	*greater or equal*
<=	a <= b	*less than or equal*

Shift. The result is the same width as **a**. The result bits are zero-filled corresponding to the size of the shift value **b**.

Operator	Context	Function
<<	a << b	*logical shift left*
>>	a >> b	*logical shift right*

8.4.1.2 Unary operators

Bit-wise. The result is the same width as **a**.

Operator	Context	Function
~	~a	*invert a*

Logical. The operand **a** must be one bit wide, and the result is one bit wide.

Operator	Context	Function
!	! a	*logical not*

Reduction. After performing the logic function on all of the bits of **a**, the result is one bit wide.

Operator	Context	Function
^	^a	*parity of a*
~^	~^a	*not parity of a*
&	&a	*reduction and*
I	I a	*reduction or*
~&	~&a	*reduction nand*
~I	~I a	*reduction nor*

8.4.1.3 Miscellaneous operators

Conditional expression. The operand **a** must be one bit wide, the operands **b** and **c** must be the same width, and the result is the same width as **b** and **c**.

Operator	Context	Example
? :	a ? b : c	**assign x = (m[3:0] == 4'd5) ? q[7:4] : r[3:0];**

Concatenation. There are no restrictions on the widths of operands **a**, **b**, and **c**, The result width is equal to the sum of the widths of the operands. The number of operands can be 2, 3, 4 etc.

Operator	Context	Example
{,}	{a,b,c}	**assign t[7:0] = {3'h2,m[0],q[5:4], r[1:0]};**

Replication. Operand **a** specifies the result width, and must be a decimal number. Operand **b** must be 1-bit wide.

Operator	Context	Example
{{}}	{a{b}}	**assign** v[7:0] = {8{t[7]}};

8.4.2 Operator precedence

In the absence of parentheses, there is a defined expression evaluation order for Verilog expression operators.

However, human memories are fallible, and verification tool implementors are imperfect. Because the writers, the readers, and the verification tools occasionally (but too often) get the operator-precedence-based evaluation order wrong, designers must ALWAYS use parenthesis in their Verilog to document and enforce the evaluation order that they intend.

An example of inconsistency with vendor tools and expression operator precedence is mixing the binary and unary "or" operators.

Example 8-9

```
wire c, a;
wire [7:0] b;
assign c = a | | b; // is not treated consistently
assign c = a | ( | b ); // is treated consistently
```

8.5 Procedural Blocks

A reader could envisage describing most or all of the logic for a design using assign statements. However, Verilog provides procedural block constructs for more elegantly expressing flip-flops, state machines and memories.

A procedural block may consist of a single procedural statement, or one or more procedural statements enclosed within **begin** and **end** keywords.

Procedural blocks may be nested within procedural blocks. The outermost procedural block must be preceded by an **always** statement or enclosed within **function** and **endfunction** keywords.

8.5.1 Combinational Logic

8.5.1.1 Procedural Assignments

Procedural assignment statements are similar to **assign** statements, except that:

- they don't have the keyword **assign** in front of them.
- their assignment target must be declared as type **reg** instead of **wire**.
- they evaluate based on changes to the signals in the **always** statement that precede them. (**assign** statements evaluate in response to changes to the signals in the expression to the right of the equal sign).

The following example illustrates combinational logic in the form of a procedural block that is functionally the same as the logic described using **assign** statements in a hierarchy of modules in section 8.3.

Example 8-10.

```
module b (p, w, x, y, z);
    input [7:0] w, x, y, z;
    output [7:0] p;
    wire [7:0] w, x, y, z;
    reg [7:0] p, r, s;
    always @(w or x or y or z)
      begin
        r = w | x;
        s = y | z;
        p = r & s;
      end
endmodule
```

There are two fundamental rules about describing combinational logic as procedural assignments.

1. The list of signals in the **always** sensitivity list must contain all the primary inputs and only the primary inputs to the combinational logic within the procedural block. For example

    - Omitting " **or** z" is incorrect. In an event-driven simulator, changes to the signal z would not trigger re-evaluation of the procedural assignments. A cycle-based simulator might use the wrong order of evaluation with respect the logic that assigns the variable z.

    - Adding " **or** s" is also incorrect. In an event-driven simulator, changes in the logic value of s would trigger unnecessary evaluations of the procedural block. For cycle-based simulator, the appearance of feedback precludes compilation.

2. The sequence of assignments must be in ascending logic rank order. In the preceding example, the assignments to r and s can be in either order, because they are at the same rank. But they must precede the assignment to r, where the expression r & s counts on updated values.

For logic in the preceding example, we could throw out the r and s intermediate variables, and the **begin** - **end** to get the following functionally equivalent logic.

Example 8-11

always @(w **or** x **or** y **or** z)
p = (w | x) & (y | z);

Note that this simplified procedural block is logically equivalent to the following **assign** statement.

   assign p = (w | x) & (y | z);

Always remember to change the type declaration for the assignment target from **reg** to **wire** when changing logic from a procedural assignment to an **assign**, and from **wire** to **reg** when changing from an **assign** to a procedural assignment.

8.5.1.2 Functions

Functions specify combinational logic that evaluates input signal values to produce an output value. The function c_data_in in [Example 8-12] concatenates an error correcting code to the left end of a data path.

Example 8-12

```
module b_dp(ap_in0,ap_in1, ap_out0,ap_out1);
    input [39:0] ap_in0, ap_in1;
    output [47:0] ap_out0, ap_out1;
    reg [47:0] ap_out0, ap_out1;
function [47:0] c_ecc_out;
    input [39:0] c_data_in;
    begin
    c_ecc_out = { (^ (c_data_in & 40'h00000007f8)),
              (^ (c_data_in & 40'h003ffff893)),
              (^ (c_data_in & 40'h0fc03f5c06)),
              (^ (c_data_in & 40'h71c3cfa84d)),
              (^ (c_data_in & 40'hb6445533ff)),
              (^ (c_data_in & 40'hd298e2fa38)),
              (^ (c_data_in & 40'h0f294bc17f)),
              (^ (c_data_in & 40'hea360cfc67)) ,
              c_data_in};
    end
endfunction
always @(ap_in0 or ap_in1)
    begin
    ap_out0 = c_ecc_out (ap_in0);
    ap_out1 = c_ecc_out (ap_in1);
    end
endmodule
```

Functions may be called from expressions in procedural statements, as in the example, and from **assign** statements. If we declare the outs in the example as wire instead of reg, we can replace the **always ... begin ... end** with

```
    assign ap_out0 = c_ecc_out (ap_in0);
    assign ap_out1 = c_ecc_out (ap_in1);
```

8.5.1.3 if-else Statement

The **if-else** keywords specify logic representing a multiplexer, as shown in [Figure 8-6].

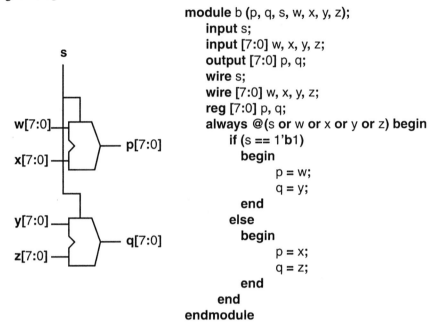

```
module b (p, q, s, w, x, y, z);
    input s;
    input [7:0] w, x, y, z;
    output [7:0] p, q;
    wire s;
    wire [7:0] w, x, y, z;
    reg [7:0] p, q;
    always @(s or w or x or y or z) begin
        if (s == 1'b1)
            begin
                p = w;
                q = y;
            end
        else
            begin
                p = x;
                q = z;
            end
    end
endmodule
```

Figure 8-6. Multiplexer represented by **if-else**

Both branches of the if-else must assign values to the same signals.

8.5.1.4 case, casex Statements

The **case** statements represent N-way multiplexers or simple state machines, while **casex** statements represent priority, one-hot or other complex encodings of state transitions.

Both statements compare the control signal within the parenthesis following the **case** or **casex** keyword with the literals preceding a colon. Where the control signal value first matches a literal, a simulator executes the statement or procedural block following the colon, and skips comparisons of the remaining literals.

The following rules must be observed in **case** and **casex** statements.

- The bit-width of the control signal must be the same as all of the literals to the left of the colons.

- Either the literal list must account for all possible values of the control signal, or

  - there must be a default keyword as the last branch preceding the endcase,

  - or a default value must be assigned immediately preceding the case statement.

The examples below illustrate these various uses of **case** and **casex** statements.

[Example 8-13] shows a 4-way multiplexer. Note that inputs w, x, y, z and s are in the **always** sensitivity list, and depending on the select input s, each one of the **case** branches assigns one of the data path input signal values to p.

Example 8-13

```
module b_mux (p, s, w, x, y, z);
    input [1:0] s;
    input [7:0] w, x, y, z;
    output [7:0] p;
    wire [1:0] s;
    wire [7:0] w, x, y, z;
    reg [7:0] p;
    always @(s or w or x or y or z)
      case (s)
        2'b00 : p = w;
        2'b01 : p = x;
        2'b10 : p = y;
        2'b11 : p = z;
      endcase
endmodule
```

This next [Example 8-14] is a state-machine implementation of a gray-code counter.

Example 8-14.
```
module b_gc (ck ,rst , r_gc);
    input ck ,rst;
    output [1:0] r_gc;
    wire [1:0] r_gc;
    reg [1:0] n;
    always @(r_gc)
      case (r_gc)
        2'b00 : n = 2'b01;
        2'b01 : n = 2'b11;
        2'b11 : n = 2'b10;
        2'b10 : n = 2'b00;
      endcase
    dff_2 reg_r ( .q(r_gc), .ck(ck), .d(n), .rst (rst) );
endmodule
```

The following [Example 8-15] shows how a **casex** can implement a priority encoder.

Example 8-15
```
module enum_encode (c_error_vector, c_code);
input   [4:0] c_error_vector;
output  [2:0]  c_code;
reg     [2:0]  c_code;
always @ (c_error_vector)
  begin
    casex (c_error_vector)
      5'b1????: c_code = 3'h1;
      5'b01???: c_code = 3'h2;
      5'b001??: c_code = 3'h3;
      5'b0001?: c_code = 3'h4;
      5'b00001: c_code = 3'h5;
      5'b00000: c_code = 3'h0;
    endcase
  end
endmodule
```

8.5.2 Storage Elements

8.5.2.1 Flip-flops

In [Example 8-14] we showed a grey-code counter with the successive counter values stored in a two-bit flip-flop submodule. [Example 8-16] shows the Verilog for that flip-flop.

Example 8-16

```
module dff_2 (q, ck ,rst , d);
    input ck ,rst;
    input [1:0] d;
    output [1:0] q;
    reg [1:0] q;
    always @(posedge ck)
        q <= (rst == 1'b0) ?d : 2'b00;
endmodule
```

8.5.2.2 Latches

Design projects nowadays generally rule out latches, except in specific areas. Some specific areas where latches may find wide use are in level-sensitive scan design (LSSD) and memory arrays. In these areas, projects generally constrain designers to use the latches in a narrow and exactly specified manner.

Latches have the design advantages of using less area than flip-flops, and providing timing performance improvements. The disadvantages are latches can complicate timing verification, as well as function verification using cycle-based simulation in many cases.

8.5.2.3 Memories

Memories are extensions of **reg** signals, with an added specification of the size of the memory following the memory name. The following line declares a memory **core**, consisting of 512 16-bit words.

reg [15:0] core [0:511];

[Example 8-17] shows a simulation model for a latch-based memory with asynchronous read.

Example 8-17.

```verilog
module m16x512(clk,we,wr_ad,rd_ad,di,r_do);
    input clk, we;
    input [7:0] wr_ad;
    input [7:0] rd_ad;
    input [15:0] di;
    output [15:0] r_do;
    reg [7:0] r_wr_ad;
    reg [7:0] r_rd_ad;
    reg r_we;
    reg [15:0] r_di;
    reg [15:0] r_do;
    reg [15:0] core [0:511];
    reg [15:0] c_do;
always @(posedge clk) begin
    r_wr_ad <= wr_ad;
    r_rd_ad <= rd_ad;
    r_we <= we;
    r_di <= di;
    r_do <= c_do;
end // always
assign c_do = core[r_rd_ad];
always @(negedge clk)
    if (r_we)
        core[r_wr_ad] = r_di;
endmodule // m16x512
```

References and assignments to memory words must specify an address, and cannot specify a subrange of a memory word. When you read or write to a memory word, you get or put an entire word.

8.5.3 Debugging

The **$display**, **$write** and **$finish** Verilog language elements for debugging described in the following sections form the starting points for addressing design problems. Techniques described in chapter 2 show how to put Verilog language elements together into an event monitor and assertion checking methodology.

There is much more to know about debugging tools and techniques that is beyond the scope of this book. There are powerful tools for debugging outside the language for detecting and diagnosing design problems. See your simulation vendor about the latest in advanced tools for finding bugs.

8.5.3.1 $display and $write Statements

The **$display** and **$write** statements consists of the keyword followed by parenthesis enclosing:

- text and optional format specifications in quotes,

- optional signals and function calls.

$display and **$write** statements are the same, except that the **$display** statement automatically inserts a newline **\n** escape character. The following two lines produce equivalent output.

> **$display("Hello Verilog");**
> **$write("Hello Verilog\n");**

Additional escape sequences for printing special characters in text include:

\t	The tab character
****	The \ character
\"	The " character
\ddd	ASCII character specified by octal **ddd** digits
%%	The % character
%m	Hierarchical path to module containing **$display** and **$write** statement

Verilog provides the following format specification options:

%b	signal in binary
%d	signal in decimal
%h	signal in hexadecimal
%o	signal in octal
%s	signal as string
%t	**$time** in decimal

Assigning 16'h7071 value to a 16-bit signal named h, the following **$display** statement:

> **$display(" t %t B %b D %d H %h o %o s %s ",$time,h,h,h,h,h);**

results in the following output line from simulation:

> t   0 B 0111000001110001 D 28785 H 7071 o 070161 s pq

8.5.3.2 $finish

When encountered during simulation, the **$finish** statement stops and exits the simulation program. Designers use **$finish** statements in testbench Verilog to stop at the end of a test, or in assertion checkers within chip design, as shown in [Example 8-18].

Example 8-18

```
...
if (c_q1_full & c_q2_full) begin
    $display (" %m c_q1_full & c_q2_full ");
    $finish;
end
...
```

8.6 Testbench

None of the examples previously presented in this chapter will do anything in simulation unless we add a testbench. Here are six components that we include in a testbench example:

- timing control,
- input stimulus,
- device under test,
- reference model,
- diagnostic logging, and
- assertion checking.

Timing control. For best results with cycle-based simulation and logic emulators, as well as preventing races in logic simulations, it is important that testbench designers isolate *timing control* into separate modules. At the core of the timing control is a master clock ck in the form shown in [Example 8-19]. The delay value #5 assigned to constant CK_MSTR specifies one half-cycle of the clock period.

Example 8-19

```
'define CK_MSTR #5
module ck_gen (ck);
    output ck;
    reg ck;
    always 'CK_MSTR ck = ~ck;
    initial ck = 1'b0;
endmodule
```

Input stimulation. [Example 8-20] illustrates a test bench input stimulus module. With each clock, it supplies successive value for the test vector r_stimulus. The module stops the simulation when it completes traversal of the complete set of r_stimulus values.

Example 8-20

```
module dff_6 (q, ck , d);
    input ck;
    input [5:0] d;
    output [5:0] q;
    reg [5:0] q;
    always @(posedge ck)
      q <= d ;
    initial
      q = 6'd0;
endmodule
module stimulus(ck,c_stimulus);
    input ck;
    output [4:0] c_stimulus;
    reg [4:0] c_stimulus;
    reg [5:0] c_counter;
    wire [5:0] r_counter;
    always @(r_counter) begin
      c_counter = r_counter + 6'd1;
      if (r_counter == 6'h20)
        $finish;
      c_stimulus = c_counter[4:0];
    end
    dff_6 reg_counter (r_counter,ck,c_counter);
endmodule
```

Device under test. The *device under test* for the testbench described in this section is the **casex** state machine enum_encode from [Example 8-15].

Reference model. To provide a *reference model* for comparison with the model being tested, we add a "enum_encodez z" module instance. It is the same as the model from [Example 8-15], except the module name has a z suffix, and it uses **casez** in place of the **casex**. This testbench verifies that a **casex** is functionally equivalent to a **casez** for all two-state values of a five-bit input.

Diagnostic logging. [Example 8-21] is a *diagnostic logging* module that tracks the progress of a test through each clock cycle. It provides output that assures the verification engineer that the stimulus module traverses the sequence of test values expected, and supports diagnosis of design errors by showing the output values side-by-side.

Example 8-21

```
'define DELAY_LOGGING #1;
module log_xz_test(ck,c_stimulus,c_codex,c_codez);
   input ck;
   input [4:0] c_stimulus;
   input [2:0] c_codex,c_codez;
   always @(posedge ck) begin
      'DELAY_LOGGING
      $display (" %t %b %h %h", $time,c_stimulus,c_codex,c_codez);
   end
endmodule
```

The DELAY_LOGGING allows designers to see the logged signal values after ck triggers updates to the registers, and their new values propagate to the logged signals.

Assertion checking. [Example 8-22] presents a simulation *assertion checker* module. It illustrates some of the simulation assertion checker concepts presented in Chapter 2, and assertion checkers as object-oriented modules discussed in Chapter 3. The DELAY_ASSERT schedules the assertion check after the last line of logged register values, the register values that trigger the assertion failure are visible.

Example 8-22

```
'define DELAY_ASSERT #2;
module assert_always (ck, event_trig_1, test, event_trig_2);
   input ck, event_trig_1, test, event_trig_2;
reg test_state;
initial test_state=1'b0;
always @(event_trig_1 or event_trig_2)
   if (event_trig_2 II event_trig_1)
      test_state = (~event_trig_2) && (event_trig_1 II test_state);
always @(posedge ck) begin
   'DELAY_ASSERT
   if((test_state==1'b1) && (test!=1'b1)) begin
      $display("ASSERTION ERROR %t:%m", $time);
      $finish;
   end
end
endmodule
```

Completing the testbench. In [Example 8-23], we put together the top-level testbench from instances of the clock *timing control* module from [Example 8-19], the stimulus *input stimulus* module in [Example 8-20], the *device under test* from [Example 8-15], and the *reference model* derived from editing [Example 8-15]. (The changes are replacing the **casex** with a **casez**, and suffixing the module name with a z.) For checking the results, the test bench instantiates log_xz_test *diagnostic logging* module from [Example 8-21], the assert_always *assertion checking* module from [Example 8-22].

Example 8-23

```
'define EVENT1 1'b1
'define EVENT2 1'b0
module testbench();
    wire ck;
    wire [4:0] c_stimulus;
    wire [2:0] c_codex;
    wire [2:0] c_codez;
    ck_gen ck_gen (ck);
    stimulus s (.ck (ck),
                .c_stimulus (c_stimulus));
    enum_encode x (.c_error_vector (c_stimulus),
                .c_code (c_codex));
    enum_encodez z (.c_error_vector (c_stimulus),
                .c_code (c_codez));
// result checking:
    log_xz_test xz_test(ck,c_stimulus,c_codex,c_codez);
    assert_always safety(ck,'EVENT1,c_codex == c_codez,'EVENT2);
endmodule
```

System design projects apply testbenches to modules, ensembles of modules in design blocks, multiple blocks in chips, and multiple chips in the system. By placing all of the clock generator, logging and assertion modules in their own separate libraries, the project benefits from the *Object-Oriented Hardware Design Principle.* Verification engineers can continue to refine the clocking, logging and assertion checking throughout the duration of the project. Likely areas of refinement include:

* Timing relationships of the clock generator(s), logging and assertion checking.

* Reduced logging through compiler code inclusion controls.

* Formatting, compression and analysis of logging and assertion outputs.

* Accommodation of cycle-based simulation and emulation.

8.7 Verilog Compilation

8.7.1 Compiler directives

Verilog compiler directives support named constant values and code inclusion controls.

8.7.1.1 Constants

Named constant value examples are state machine state names, data path fields/widths, and memory array sizes.

Here we have state machine state names defined in terms of numeric equivalents.

```
'define    R_NORMAL    2'h0
'define    R_WAIT      2'h1
'define    R_DONE      2'h2
'define    R_IDLE      2'h3
```

The following lines define names for bit fields.

```
'define    IO_T_FLD    15:10
'define    IO_Q_FLD    9:4
'define    IO_S_FLD    3:0
```

This line defines a memory array size.

```
'define    QT_DEPTH    64
```

Using 'include directives, a design project can centralize their 'define constant definitions on a few files, and share the definitions across all the modules that use these constant definitions. Wherever one module writes a code value to a signal or memory and another module must read that same information, use a 'include pointing to the same file of 'define values in both modules.

The following [Example 8-24] illustrates an application of the grey code to the state machine states from a file named rtype.h.

Example 8-24

```
module b_gc (ck ,rst , r_gc);
'include "rtype.h"
    input ck ,rst;
    output [1:0] r_gc;
    wire [1:0] r_gc;
    reg [1:0] n;
    always @(r_gc)
      case (r_gc)
        'R_NORMAL : n = 'R_WAIT;
        'R_WAIT : n = 'R_IDLE;
        'R_IDLE : n = 'R_DONE;
        'R_DONE : n = 'R_NORMAL;
      endcase
    dff_2 reg_r ( .q(r_gc), .ck(ck), .d(n), .rst (rst) );
endmodule
```

Constant expressions are useful when designers specify the width of signals that hold ASCII 8-bit codes, or memory arrays in terms of the total word capacity.

```
reg [(8 * 'SWDTH) - 1 ) : 0] r_chip_id;
reg ['IO_T_FLD] qt_array [0: 'QT_DEPTH-1];
```

8.7.1.2 Code Inclusion

Code inclusion controls provide a design project with a mechanism through which they can compile a single body of source files optimally targeting selected simulation goals, such as detection, diagnosis, or coverage measurement.

Selection of a code inclusion option begins with the presence or absence of the option name following **+define+** on the simulation compile command line.

<vendor-compile-command> **+define+**RECORDOFF *<compile-options>*

Then, in the Verilog for the test bench, chips, and libraries, the compiler tests the **'ifdef**-**'else**-**'endif** sequences.

```
'ifdef 'RECORDOFF
'else
<trace-file recording PLI calls>
'endif
```

8.7.1.3 Command Line

Compilation. Many Verilog EDA tool vendors support the following compile-time command line options:

+define+<*name*> **+incdir+**<*directory path*> <*files*> **-v** <*library-file*> |
-f <*command-file*> **-F** <*relative-command-file*>

where:

+define+<*name*>	compile time code inclusion option names
+incdir+<*directory path*>	paths to directories containing files of constants for '**include** "<*file-name*>" in the Verilog files and <*library-files*>
<files>	list of files containing the Verilog code for the modules in a design.
-v <*library-file*>	Verilog compilers only process the modules from these files as they are needed in the design module files.
-f <*command-file*>	file containing compile options. Relative <*files*> in the **-f** <*command-file*> are relative to the directory where the command line executes.
-F <*command-file*>	file containing compile options. Relative. <*files*> in the **-F** <*command-file*> are relative to the directory where the <*command-file*> resides.

Verilog compilers read the entire command line and all of the command files before reading the files specified. Constants and include directories defined at the end of the options are visible to all of the Verilog files and libraries.

Compile options specified on the command line and in command files are the same. The layout of the options across lines is different.

- Continuation of a command line requires a backslash \ that tells the command line parser to go on to the next line for more options. An end-of-line with no \ marks the end of options on a command line.

- Command files do not use the backslash for continuation across lines. The command file parser treats the end-of-line as white space separating options. The end-of-file marks the end of options on a command file.

- Command files allow use of Verilog //-to-end-of-line comments.

Simulation. Simulation run options widely supported by vendor simulators include:

     -l *<log_file>* **+** *<user-plusarg>*

where:

    -l *<log_file>*   Simulator writes **$display** and **$write** messages to the specified *<log_file>* name.

    + *<user-plusarg>*   Provides simulation run controls.

The Verilog **$test$plusargs** (*<user-plusarg>*) built-in function tests for +*<user-plusarg>* in the command line.

The Verilog PLI **mc_scan_plusarg** (*<user-plusarg>*) function tests for +*<user-plusarg>* and returns the remainder of the string that follows the *<user-plusarg>*. A simulation command line argument of the form:

    +rseed=89674523

can be detected and processed by the following PLI C code.

```
char    *plusarg;
...
plusarg = mc_scan_plusargs("rseed=");
```

After executing the function call, plusarg contains a pointer to a string containing "89674523."

8.8 Summary

This chapter presented a tutorial on Verilog language elements applicable register transfer abstraction level and their verifiable use.

For verifiability, we emphasized strong typing and fully-specified state machines using **case**, **casex** and **if-else** statements. Since the Verilog X-state is counter-productive in RTL verification (see chapter 7), we omitted it.

We reviewed debugging statements, constant naming, code inclusion controls and command line options for compilation and simulation in a verification environment.

9

Principles of Verifiable RTL Design

The conception of a verifiable RTL philosophy is a product of two factors: one, inherited seat-of-the-pants experiences during the course of large system design; the other, the sort of investigation which may be called "scientific." Our philosophy falls somewhere between the knowledge gained through experiences and the knowledge gained through scientific research. It corroborates on matters as to which definite knowledge has, so far, been ascertained; but like science, it appeals to reason rather than authority. Our philosophy consists of a fundamental set of principles, which when embraced, yield significant pay back during the process of verification.

9.1 Principles

In general, a principle is a comprehensive and fundamental law or doctrine--while a set of principles constitutes a philosophy. In this section we summarize the principles which combine to form the framework for our verifiable RTL philosophy.

9.1.1 Disciplined User Principle

The *Disciplined User Principle*, introduced in Chapter 1, states that designers who limit their degrees of freedom in writing RTL will encounter the least anomalies in tool behavior. A disciplined design approach provides cooperation with the various tools used during the process of design and verification. This cooperation can return an order of magnitude improvement in performance and capacity of simulation and equivalence checking tools--while providing a common mechanism for communicating design intent across a design organization.

9.1.2 Fundamental Verification Principle

The *Fundamental Verification Principle,* introduced in Chapter 2, states that the implementing of RTL code must follow completion of the specification to avoid unnecessarily complex and unverifiable designs. Developing an unambiguous specification is fundamental to verifiable design.

9.1.3 Retain Useful Information Principle

The *Retain Useful Information Principle,* introduced in Chapter 2, states that a single process within a design flow should never discard information that a different process within the flow must reconstruct at a significant cost. To conveniently operate between the RTL description and the physical flow, it is important that the various transformations consider subsequent process requirements within the flow. An example of applying this principle would be embedding the hierarchical RTL signal and wire names in the physical design during flattening, which will be used by the equivalence checking process for cutpoint identification. An alternative example is the use of assertion checkers and event monitors for capturing design assumptions, environmental constraints, and expected behavior during the RTL implementation phase. The loss of this design knowledge and environmental assumptions can result in both higher verification and maintenance costs.

9.1.4 Orthogonal Verification Principle

The *Orthogonal Verification Principle,* introduced in Chapter 2, states that functional behavior, logical equivalence and physical characteristics should be treated as orthogonal verification processes within a design flow. This principle provides the foundation for today's static verification design flows, which enables a verification process to focus on its appropriate concern through abstraction. By applying this principle, we are able to achieve orders of magnitude faster verification, support larger design capacity, and higher verification coverage including exhaustive equivalence and timing analysis.

9.1.5 Functional Observation Principle

The Functional Observation Principle, introduced in Chapter 2, state that a methodology must be established which provides a mechanism for observing and measuring specified function behavior. Without looking for specific events and assertions during the course of verification, the designer has no convenient means for measuring functional correctness. Chapter 2 describes a technique of combining event monitors, assertion checkers and coverage tools to form a methodology which satisfy the Functional Observation Principle.

9.1.6 Verifiable Subset Principle

The *Verifiable Subset Principle*, introduced in Chapter 3, states that a design project must select a simple HDL *verifiable subset*, which serves all verification tools within the design flow as well as providing an uncomplicated mechanism for conveying clear functional intent between designers. [Table 9.1] summarizes our recommended Verilog RTL verifiable subset.

Table 9.1 Verifiable Subset

always	else	initial	parameter
assign	end	inout	posedge
begin	endcase	input	reg
case	endfunction	module	tri
casex	endmodule	negedge	tri0
default	function	or	tri1
defparam	if	output	wire

To achieve a verifiable RTL design, we believe it necessary to adopt and enforce a discipline in coding style and RTL subset.

9.1.7 Object-Oriented Hardware Design Principle

The *Object-Oriented Hardware Design Principle*, introduced in Chapter 3, states that design engineers must code at a higher object level of abstraction--as oppose to a lower implementation or tool specific detailed level--to facilitate verification process optimizations and design reuse. By applying the *principle of information hiding* on the functional grouping of state elements (and other objects) within the RTL, and introducing a new level of design abstraction, the design engineer will succeed in isolating design details within tool-specific libraries. This methodology allows for the simultaneous optimization of simulation, equivalence-checking, model-checking and physical design within the design flow. Furthermore, this methodology allows for the augmentation of new features and tools throughout the duration of a project, without interfering with the text or functional intent of the original design.

9.1.8 Project Linting Principle

The *Project Linting Principle*, introduced in Chapter 3, states that project specific coding rules must be enforced automatically within the design flow to ensure productive use of design and analysis tools, as well as improving communication between design engineers. A linting methodology must be established early in the design cycle, and used to enforce all verifiable subset and project specific coding style requirements and rules. Ideally, the linting process should be embedded directly into the design flow (in Makefiles, for example), and used to prevent advancing to subsequent processes within the flow upon detection of errors or code style violations. Enforcing a project-specific coding style allows us to achieve a truly verifiable RTL design and is key to our verifiable RTL philosophy.

9.1.9 Fast Simulation Principle

The *Fast Simulation Principle*, introduced in Chapter 4, states that a design project must tailor its RTL (and its design process) to achieve the fastest simulation possible. An underlying value that pervades this book is the importance of fast simulation, particularly at later stages in the project, when bugs are few and far between. Fast simulation at this phase of design offers the following benefits: (a) reduces the bug detection rate sooner, (b) provides more productive use of the CPU's in the simulation farm, and (c) potentially locates a few more of those far-between bugs before silicon.

9.1.10 Visit Minimization Principle

The *Visit Minimization Principle*, introduced in Chapter 4, states that for best simulation tool performance, minimize the frequency and granularity of procedural block, signal, and bit visits. To get the best performance from logic simulation, some understanding of the way that simulators work is desirable. Fundamental to logic simulation performance is minimizing the number of visits.

9.1.11 Cutpoint Identification Principle

The *Cutpoint Identification Principle*, introduced in Chapter 5, states that a single design decision pertaining to functional complexity must be isolated and localized within a module to facilitate equivalence checking cutpoint identification. Isolating the functional complexity of multipliers is a classic example of how the equivalence checker's runtime performance can be improved when applying this principle.

9.1.12 Test Expression Observability Principle

The *Test Expression Observability Principle*, introduced in Chapter 5, states that a complex test expression within a case or if statement must be factored into a variable assignment. Complex test expressions within a Verilog case statement can complicate the verification process. It is easier to debug the branching effect within a simulation trace file when the case test expression is an observable variable. In addition, equivalence checking the RTL description against a gate level implementation is improved through the addition of potential cutpoints.

9.1.13 Numeric Value Parameterization Principle

The *Numeric Value Parameterization Principle*, introduced in Chapter 5, states that numeric values should be parameterized and not directly hard-coded into the RTL source. For example queue structures must be treated as independent objects and abstracted away from other functionality within the Verilog RTL. This is achieved by applying the Numeric Value Parameterization Principle. In other words, a parameterized model for the queue should be instantiated to provide a mechanism for queue size reduction during verification. Queue depth and word size reduction will result, in many instances, in an improvement in model checking runtime performance while potentially preventing the condition of state explosion.

9.1.14 Indentation Principle

The *Indentation Principle*, introduced in Chapter 6, states that a design project must define a uniform indentation policy. Designers writing Verilog source code who follow source line indentation policies help other engineers reading the Verilog code to see the relationship of control statements. The authors believe that designers must adopt the habits that support reuse, and use spaces instead of tabs even for modules in projects that have no possible use outside the project domain.

9.1.15 Meta-comment Principle

The *Meta-comment Principle*, introduced in Chapter 6, states that vendor-specific meta-comments must be avoided whenever possible. In general, it is preferable to use standard, non-proprietary and vendor-independent meta-comments (or other methods) to specify extra application directives.

Use of vendor-neutral source for a design allows a project maximum flexibility in its tool choices, and may facilitate reuse on future projects.

9.1.16 Asynchronous Principle

The *Asynchronous Principle*, introduced in Chapter 6, states that a design project must minimize and isolate resynchronization logic between asynchronous clock domains. It is of the utmost importance for design project members to understand that an increased resynchronization frequency (or a slower resync circuit speed) can result in a catastrophic change in device failure rates. Failure rates can go from one every million years to three per day by doubling the frequency.

9.1.17 Combinational Feedback Principle

The *Combinational Feedback Principle*, introduced in Chapter 6, states that designers must not use any form of combinational logic feedback (real, false-path, or apparent) in their Verilog. In modern logic design practice, combinational logic feedback is universally avoided. Verification tools that count on no combinational logic feedback (cycle-based simulators, equivalence checkers, timing verifiers) diagnose such feedback loops. Designers occasionally (though rare) will inadvertently specify combinational feedback loops in their RTL. Three sources of feedback loops that can hinder the RTL verification process are design errors, false paths and apparent (not real) feedback.

9.1.18 Code Inclusion Control Principle

The *Code Inclusion Control Principle*, introduced in Chapter 6, states that a design project must define and document code inclusion controls (e.g. 'ifdef IDENTIFIER) and provide a process for managing them. On large design projects, the Verilog files for a design are the work product of a large number of engineers, so there is a danger that code inclusion controls may become complicated and redundant if their definition and use are not coordinated.

9.1.19 Entry Point Naming Principle

The *Entry Point Naming Principle*, introduced in Chapter 6, states that RTL tool libraries must support a prefix-based entry point naming convention. Identical module names, user task and function names conflict in the global name space. For the reasons discussed in 6.4.1.4, user tasks and function naming must consider simulation performance profiling and module integration into a system model. The authors recommend project-wide allocation of prefixes to user tasks and function names, as well as their corresponding PLI function library.

9.1.20 Faithful Semantics Principle

The *Faithful Semantics Principle*, introduced in Chapter 7, states that an RTL coding style and set of tool directives must be selected which insures semantic consistency between simulation, synthesis and formal verification tools. To avoid RTL and gate-level simulation differences, design projects should adopt the RTL Verilog style presented in this book. They must enforce a project specific style by tailoring a linting tool rules set, and locking the linting step into their design process to check all RTL Verilog.

When a project does not enforce faithful semantics, RTL simulations loses its credibility, and significantly more gate-level simulation is required. Because equivalence checkers base their RTL semantics on synthesis RTL policies, they are generally no help in detecting RTL and synthesized gate simulation semantic differences.

9.1.21 Good Vendor Principle

The *Good Vendor Principle*, introduced in Chapter 7, states that verification tool vendors must support real user needs in a project's design environment, not the tool vendor's preferred environment. Design projects increasingly rely on EDA vendor tools for their success in design verification. In addition to contributing to the success in verification on design projects, the EDA vendor verification tools too often add difficulties to a project's verification process. It is unfair, however, to ask a vendor to comply with the *Good Vendor Principle* when a design project does not comply with the *Disciplined User Principle*. The two principles complement each other.

9.2 Summary

In the early 1990s, the system design communities underwent tremendous productivity gains in gate-level design as engineers embraced synthesis technology. Unfortunately, this resulted in an increase in the design's verification problem space for the design as well as the verification process. To keep up with escalating design complexity and sizes, we have presented a Verilog RTL coding style and a verifiable subset that facilitates optimizing the verification flow. We have emphasized the importance of two-state simulation, which we believe is fundamental to the RT level verification process--particularly at identifying start-up state initialization problems.

A verifiable RTL coding methodology permits the engineer to achieve greater verification coverage in minimal time, enhances cooperation and support for multiple EDA tools within the flow, clarifies RTL design intent, and facilitates emerging verification processes. The design project will accomplish a reduction in development time-to-market while simultaneously achieving a higher level of verification confidence in the final product through the adoption a Verifiable RTL design methodology.

Bibliography

[Abramovici *et al* 1990] M. Abramovici, M. A. Breuer, A. D. Friedman, *Digital Systems Testing and Testable Design*, IEEE Press, New York.

[Abts 1999] D. Abts, "Integrating Code Coverage Analysis into a large Scale ASIC Design Verification Flow," *Proc. Intn'l HDL Conferenc*e, pp. 141-145, April, 1999.

[Amdahl 1967] G. M. Amdahl, "Validity of the Single Processor Approach to Achieving Large Scale Computing Capabilities," *Proc. Spring Joint Computer Conference*, pp. 483-485, 1967.

[Arnold *et al* 1998] M. G. Arnold, N. J. Sample, J. D. Schuler, Guidelines for Safe Simulation and Synthesis of Implicit Style Verilog, *Proc. Intn'l Verilog HDL Conferenc*e, pp. 59-66, March, 1998.

[Ashar and Malik 1995] P. Ashar and S. Malik, "Fast functional simulation using branching programs," *Proc. Intn'l Conf. on Computer-Aided Design*, pp. 408-412, 1995.

[Barbacci and Siewiorek 1973] M. B. Barbacci and D. P. Siewiorek, "Automated Exploration of the Design Space for Register Transfer (RT) Systems," *Proc. First Annual Symposium on Computer Architecture*, December, 1973.

[Barnes and Warren 1999] P. Barnes, M. Warren, "A Fast and Safe Verification Methodology Using VCS, "*Synopsys User's Group (SNUG99)*, Retrieved August 23, 1999 from the World Wide Web: *http://www.synopsys.com/news/pubs/snug/snug99_papers/Barnes_Final.pdf*

[Beizer 1990] B. Beizer, *Software Testing Techniques*, Van Nostrand Rheinhold, New York, second edition, 1990.

[Bening 1969] L. Bening, "Simulation of High Speed Computer Logic," *Proc. Design Automation Workshop*, pp. 103-112, June, 1969.

[Bening et al. 1982] L. Bening, T. A. Lane, C. R. Alexander, J. E. Smith, "Developments in Logic Network Path Delay Analysis," *Proc. Design Automation Conference*, pp. 605-615, June, 1982.

[Bening et al. 1997] L. Bening, T. Brewer, H. D. Foster, J. S. Quigley, R. A. Sussman, P. F. Vogel, and A. W. Wells, "Physical Design of 0.35µ Gate Arrays for Symmetric Multiprocessing Servers," *Hewlett-Packard Journal*, pp. 95-103, April, 1997.

[Bening 1999a] L. Bening, "An RTL Design Verification Linting Methodology," *Proc. Intn'l HDL Conference*, pp. 136-140, April, 1999.

[Bening 1999b] L. Bening, "A Two-State Methodology for RTL Logic Simulation," *Proc. Design Automation Conference*, pp. 672-677, June, 1999.

[Berman and Trevillyan 1989] C. L. Berman and L. H. Trevillyan. "Functional Comparison of Logic Designs for VLSI Circuits," *Proc. Intn'l. Conf. on Computer-Aided Design*, pp. 456-537, 1989.

[Berge' et al. 1995] J.-M. Berge', O. Levia, J. Rouillard, "High-level System Modeling Specification Languages," in *Current Issues in Electronic Modeling, Volume 3*, pp. 51-75, Kluwer Academic Publishers, 1995.

[Blank 1984] T. Blank, "A Survey of Hardware Accelerators Used in Computer-aided Design," *IEEE Design and Test*, pp. 21-39, Aug., 1984.

[Brand 1993] D. Brand, "Verification of Large Synthesized Designs," *Proc. Intn'l. Conf. on Computer-Aided Design*, pp. 534-537, 1993.

[Brayton et al. 1996] R. Brayton, G. Hachtel, A. Sangiovanni-Vincentelli, F. Somenzi, A. Aziz, S. Cheng, S. Edwards, S. Khatri, Y. Kukimoto, A. Pardo, S. Qadeer, R. Ranjan, S. Sarwary, T. Shiple, G Swamy, T. Villa, "VIS: A System for Verification and Synthesis," *Proc. Computer Aided Verification*, 1996

[Breuer 1972] M. A. Breuer, "A Note on Three-valued Logic Simulation," *IEEE Trans. on Computers*, vol. C-21, pp. 399-402, Apr., 1972.

[Bryant 1986] R. Bryant, "Graph-based Algorithms for Boolean Function Manipulation," *IEEE Trans. on Computers*, Vol. C-35, No. 8, pp.677-691.

[Buchnik and Ur 1997] E. Buchnik, S. Ur, "Compacting regression-suites on-the-fly," *Proceedings of the 4th Asia Pacific Software Engineering Conference*, 1997.

[Burch and Singhal 1998] J. R. Burch and V. Singhal, "Tight Integration of Combinational Verification Methods," *Proc. Intn'l Conf. on Computer-Aided Design*, 1998.

[Cerny and Mauras 1990] E. Cerny and C. Mauras, "Tautology Checking Using Cross-controllability and Cross-Observability Relations," *Proc. Intn'l. Conf. on Computer Aided Design*, pp. 34-37, 1990.

[Cerny et al. 1998] E. Cerny, B. Berkane, P. Girodias, K. Khordoc, *Hierarchical Annotated Action Diagrams*, Kluwer Academic Publishers, 1998.

[Chappell 1999] B. Chappell, "The Fine Art of IC Design," *IEEE Computer*, pp. 30-34, July, 1999.

[Chappell and Yau 1971] S. G. Chappell and S. S. Yau, "A Three-Value Design Verification System," *Proc. Fall Joint Computer Conference*, pp. 651-661, 1971.

[Cheng and Krishnakumar 1993] K-T. Cheng, A. Krishnakumar, "Automatic Functional test Generation Using the Extended Finite State Machine Model." *Proc. Design Automation Conference*, pp. 86-91, June, 1993.

[Chu 1965] Y. Chu, "An Algol-like Computer Design Language," *Communications of the ACM*, pp.607-615, Oct. 1965

[Clarke, Emerson and Sistla1981] E. M. Clarke, E. A. Emerson, and A. P. Sistla, "Characterizing Properties of Parallel Programs as Fixpoints." *Seventh Intn'l Colloquium on Automata, Languages, and Programming, volume 85 of LNCS*, 1981.

[Clarke and Wing 1996] E. Clarke, J. Wing, "Formal Methods: State of the Art and Future Directions," *CMU Computer Science Technical Report CMU-CS-96-178*, August 1996

[Clarke and Kurshan 1997] E. Clarke, R. Kurshan, "Computer-Aided Verification," *IEEE Spectrum*, pp.61-67, June 1997.

[Devadas et al. 1996] S. Devadas, A. Ghosh, K. Keutzer, "An Observability-Based Code Coverage Metric for Functional Simulation," *Proc. Intn'l Conf. on Computer-Aided Design*, pp. 418-425, 1996.

[Dewey 1992a] A. Dewey ed. "Three Decades of HDLs Part 1: CDL Through TI-HDL," *IEEE Design and Test*, pp. 69-81, June, 1992.

[Dewey 1992b] A. Dewey ed. "Three Decades of HDLs Part 2: Conlan Through Verilog," *IEEE Design and Test*, pp. 54-63, September, 1992.

[Dietmeyer and Duley 1975] D. Dietmeyere, J. Duley, "Register Transfer Languages and Their Translation," In M. Breuer, *Digital System Design Automation: Languages, Simulation & Data Base*, pp. 117-218, Computer Science Press, Inc. 1975.

[Dill and Tasiran 1999] D. Dill, S. Tasiran, "Simulation meets formal verification," *Proc. Intn'l Conf. on Computer-Aided Design*, pp.221, 1999. Retrieved November 21, 1999 from Stanford University database on the World Wide Web: http://verify.stanford.edu/

[Duley and Dietmeyer 1968] J. Duley, D. Dietmeyer, "A Digital System Design Language (DDL)," *IEEE Trans. on Computers*, Vol.C-17, No. 9, pp. 850-861, Sept. 1968.

[Eichelberger and Williams 1977] E. B. Eichelberger and T. W. Williams, "A logic Design Structure for LSI Testability," *Proc. Design Automation Conference*, pp. 462-468, June, 1977.

[Ellsberger 1997] J. Ellsberger, D. Hogrefe A. Sarma, *SDL: Formal Object-oriented Language for Communicating Systems*, Prentice Hall, 1997.

[Eiriksson 1996] A. Eiriksson, "Integrating Formal Verification Methods with A Conventional Project Design Flow," *Proc. Design Automation Conference*, pp. 666-671, 1996.

[Fallah et al. 1998] F. Fallah, S. Devadas, K. Keutzer, "OCCOM: Efficient Computation of Observability-Based Code Coverage Metrics for Functional Verification," *Proc. Design Automation Conference*, pp.152-157, 1998.

[Foster 1998] H. D. Foster, "Techniques for Higher Performance Boolean Equivalence Verification," *Hewlett-Packard Journal*, pp. 30-38, August, 1998. http://www.hp.com/hpj/98aug/au98a3.htm

[Foster 1999] H. Foster "Formal Verification of the Hewlett-Packard V-Class Servers," *Proc. DesignCon99 On-Chip Design Conference*, January, 1999, pp. 107-119.

[Grinwald et al. 1998] R. Grinwald, E. Harel, M. Orgad, S. Ur, A. Ziv, "User Defined Coverage - A tool Supported Methodology for Design Verification," *Proc. Design Automation Conference*, pp. 158-163, 1998.

[Gupta et al. 1997] A. Gupta, S. Malik, P. Ashar, "Toward Formalizing a Validation Methodology Using Simulation Coverage," *Proc. Design Automation Conference*, pp. 740-745, 1997.

[Hefferan *et al.* 1985] P. H. Hefferan, R. J. Smith, V. Burdick, D. L. Nelson, "The STE-264 Accelerated Electronic CAD System," *Proc. Design Automation Conference*, pp. 352-358, June, 1985.

[Hill and Peterson 1973] F. Hill, G. Peterson, *Digital Systems: Hardware Organization and Design*, Wiley, New York, 1973.

[Hitchcock 1982] R. B. Hitchcock, "Timing Verification and the Timing Analysis program," *Proc. Design Automation Conference*, pp. 594-604, June, 1982.

[Hoare 1981] C. A. R. Hoare, "The Emporer's Old Clothes," *Communications of the ACM*, February, 1981, pp. 75-83.

[Hoare 1998] C. A. R. Hoare, "The Logic of Engineering Design," Retrieved August 4, 1999 from the World Wide Web: http://www.comlab.ox.ac.uk/oucl/users/tony.hoare/logic1.html, March, 1998.

[Horgan et al. 1994] J. Horgan, S. London, M. Lyu, "Achieving Software Quality with Testing Coverage Measures," *Computer*, 27(9), pp. 60-69, September 1994.

[Huang and Cheng 1998] S.Y. Huang and K.T. Cheng, *Formal Equivalence Checking and Design Debugging*, Kluwer Academic Publishers, 1998.

[Hughes 1958] B. Hughes, One of Minnesota's Newest Firms - Control Data Corporation, *Minnesota Technolog*, pp. 33-36, April, 1958.

[IEEE 1076 1993] IEEE Standard 1076-1993 *VHDL Language Reference Manual*, IEEE, Inc., New York, NY, USA, June 6, 1994.

[IEEE 1364 1995] IEEE Standard 1364-1995 *IEEE Standard Hardware Description Language Based on the Verilog Hardware Description Language*, IEEE, Inc., New York, NY, USA, October 14, 1996.

[IEEE 1364.1 1999] IEEE P1364.1/D1.4, *Draft Standard for Verilog Register Transfer Level Synthesis*, IEEE, Inc., New York, NY, USA, April 26, 1999.

[Iverson 1972] K.E. Iverson, "A Common Language for Hardware, Software, and Applications," *Proceedings of the 1972 FJCC*, pp.121-129, 1972.

[Jephson *et al* 1969] J. S. Jephson, R. P. McQuarrie, R. E. Vogelsberg, "A Three-value Design Verification System," IBM Systems Journal, Vol. 8, No. 3, pp. 178-188, 1969.

[Kang and Szygenda 1992] S. Kang and S. Szygenda, "Modeling and Simulation of Design Errors," *Proc. of the Int'l Conference on Computer Design: VLSI in Computers and Processors,* pp. 443-446, October 1992.

[Kantrowitz and Noack 1996] M. Kantrowitz, L. Noack, "I'm Done Simulating; Now What? Verification Coverage Analysis and Correctness Checking of the DECchip 21164 Alpha microprocessor," *Proc. Design Automation Conference*, pp. 325-330, 1996.

[Keating and Bricaud 1999] M. Keating and P. Bricaud, *Reuse Methodology Manual*, Kluwer Academic Publishers, 1999.

[Kleeman and Cantoni 1987] L. Kleeman and A. Cantoni, Metastable Behavior in Digital Systems, *IEEE Design and Test*, pp. 4-19, Dec., 1987.

[Kleinrock 1991] Networks". Kleinrock, "ISDN - The Path to Broadband Networks." *Proc. IEEE*, Feb. 1991, pp. 112-117.

[Krohn 1981] H. E. Krohn, "Vector Coding Techniques for High Speed Simulation," *Proc. Design Automation Conference*, pp. 525-529, 1981.

[Kuehlmann and Krohm 1997] A. Kuehlmann and F. Krohm, "Equivalence Checking Using Cuts and Heaps," *Proc. Design Automation Conference*, pp. 263-268, 1997.

[Kunz 1993] W. Kunz, "HANNIBAL: An Efficient Tool for Logic Verification Based on Recursive Learning," *Proc.Intn'l Conf. on Computer-Aided Design*, pp. 538-543, 1993.

[Kurshan 1994] R. P. Kurshan, *Computer-Aided Verification of Coordinating Processes: The Automata-Theoretic Approach*, Princeton University Press, 1994.

[Kurshan 1997] R. P. Kurshan, "Formal Verification in a Commercial Setting," *Proc. Design Automation Conference*, pp. 258-262, 1997.

[Malka and Ziv 1998] Y. Malka and Avi Ziv, "Design Reliability - Estimation Through Statistical Analysis of Bug Discovery Data," *Proc. Design Automation Conference*, pp. 644-649, June, 1998.

[Mangelsdorf *et al.* 1997] S. Mangelsdorf, R. Gratias, R. Blumberg, R. Bhatia, "Functional Verification of the HP PA 8000 Processor," *Hewlett-Packard Journal*, August, 1997.

[Matsunaga 1996] Y. Matsunaga, "An Efficient Equivalence Checker for Combinatorial Circuits," *Proc, Design Automation Conference*, pp. 629-634, 1996.

[McGeer *et al.* 1995] P. C. McGeer, K. L. McMillan, A.Saldanha, A. L. Sangiovanni-Vincentelli. and P.Scaglia, P., "Fast Discrete Function Evaluation Using Decision Diagrams," *Proc. Intn'l Conf. on Computer-Aided Design*, pp. 402-407, November, 1995.

[McMillan 1993] K. L. McMillan, *Symbolic Model Checking*, Kluwer Academic Publishers, 1993.

[McWilliams 1980] T. M. McWilliams, "Verification of Timing Constraints on Large Digital Systems," *Proc, Design Automation Conference*, pp. 139-147, 1980.

[Mills and Cummings 1999] D Mills and C. Cummings. "RTL Coding Styles That Yield Simulation and Synthesis Mismatches." Synopsys Users Group, San Jose, 1999. Retrieved August 3, 1999 from Synopsys database on the World Wide Web: http://www.synopsys.com/news/pubs/snugsnug99_papers/Mills_Final.pdf

[Mittra 1999] S. Mittra, Principles of VERILOG PLI, Kluwer Academic Publishers, 1999.

[Murata 1989] T. Murata, Petri Nets: Properties, Analysis and Applications, *Proc. IEEE*, vol. 77, no. 1, pp. 541-580, Apr. 1989.

[OVI LRM 1993] Open Verilog International, *Verilog Hardware Description Language Reference Manual (LRM) Version 2.0*, March, 1993.

[Parnas 1972] D. L. Parnas, "On the Criteria to be Used in Decomposing Systems Into Modules," *Communications of the ACM*, Vol. 5, No 12, pp. 1053-1058, December 1972.

[Pfleeger 1998] S. L. Pfleeger, *Software Engineering: Theory and Practice*, Prentice Hall, 1998

[Ross and Goodenough 1975] D.T. Ross, J.B. Goodenough, C.A Irvin, "Software Engineering: Process, Principles, and Goals," *IEEE Computer*, Vol. 8, No. 5, pp. 17-27, May 1975.

[Rowson and Sangiovanni-Vincentelli 1997], J. Rowson, A. Sangiovanni-Vincentelli "Interface-based Design," *Proc. Design Automation Conference*, pp. 178-183, 1997.

[Sangiovanni-Vincentelli et al. 1996], A. Sangiovanni-Vincentelli, P. McGeer, A. Saldanh, "Verification of Electronic Systems," *Proc. Design Automation Conference*, pp. 106-111, 1996.

[Seshu and Freeman 1962] S. Seshu and D. M. Freeman, "The Diagnosis of Asynchronous Sequential Switching Systems," *IEEE Trans. on Elec. Computers*, Vol 11, pp. 459-465, August, 1962.

[Seshu 1965] S. Seshu, "On an Improved Diagnosis Program," *IEEE Trans. on Elec. Computers*, Vol 12, pp. 76-79, February, 1965.

[Sutherland 1999] S. Sutherland, *The Verilog PLI Handbook : A User's Guide and Comprehensive Reference on the Verilog Programming Language Interface*, Kluwer Academic Publishers, Norwell, MA 02061, 1999.

[Szygenda 1972] S. A. Szygenda, "TEGAS2 -- Anatomy of a General Purpose Test Generation and Simulation System for Digital Logic," *Proc. Design Automation Conference*, pp. 116-127, June, 1972.

[Taylor *et al.* 1998] S. Taylor, M. Quinn, D. Brown, N. Dohm, S. Hildebrandt, J. Huggins, J. and C. Ramey, "Functional Verification of a Multiple-issue Out-of-order, Superscalar Alpha Processor -- the DEC Alpha 21264 microprocessor," *Proc. Design Automation Conference*, pp. 638-643, June, 1998.

[Thomas and Moorby 1998] D. E. Thomas and P. R.Moorby, *The Verilog Hardware Description Language*, Kluwer Academic Publishers, Norwell, MA 02061, pp. 136, 4th Edition, 1998.

[Ulrich 1965] E. G. Ulrich, "Time Sequenced Logical Simulation Based on Circuit Delay and Selective Tracing of Active Network Paths," *Proc. ACM National Conference*, pp. 437-448, 1965.

[Wilcox and Rombeck 1976] P. Wilcox and H. Rombeck, "F/Logic - An Interactive Fault and Logic Simulator for Digital Circuits," *Proc, Design Automation Conference*, pp. 68-73, 1976.

A

Comparing Verilog Construct Performance

The design examples in this book only show a few of the many functionalities that designers need to describe in Verilog. When comparing the simulation performance of two possible ways of expressing the same behavior, performance numbers may vary between simulators from different vendors, and between different versions of Verilog simulators from the same vendor. When writing any high-usage functionality in Verilog, a designer should measure the performance of alternative ways for writing the Verilog, and adopt the one that simulates the fastest.

We have found that in most cases, the faster Verilog will be simpler and more clearly express the design.

Example A-1 shows a test bench for measuring simulation performance. It measures the performance of two different methods of expressing bus reversal. Lines 16-17 show bus reversal expressed as a **for** loop, and line 20 expresses the same function as concatenation. Using this test bench for performance testing proceeds in four compile and simulate steps.

1. Use **+define**+SHORT+CHECK+METHOD1+METHOD2+ to see whether the two methods have equivalent functionality.
2. Use **+define+**METHOD1+ and record the simulation time M1 for executing bus reversal as a for loop.

3. Use **+define+**METHOD2+ and record the simulation time M2 for executing bus reversal using concatenation.

4. Omit **+define+** and record the simulation time Z for executing the empty loop.

You can then determine the speedup factor S for the lines under test using the following formula.

$$S = ((M1 - Z)/(M2 - Z))$$

Here are some performance measurement and optimization factors illustrated in the example that you must consider when measuring performance.

- Small CPU run time measurements of less than a second are not very accurate. The 10 million loop iteration count accumulates a run time that gets the timing measurement up to between 1 and 100 seconds. Depending on the complexity of the Verilog construct that you are comparing, and the performance of you host computer and simulator, you may have to adjust the iteration count to get to an reasonable 1-100 second CPU time value.

- If the result calculated is never used subsequently, some optimizers will delete earlier statements that calculated the result. By accumulating exclusive or results and then **$display**ing them before the **$finish;** the example forces the optimization to retain the lines leading up to a final values for e0 and e1.

- Many compilers can recognize that an expression will not change with loop iterations, and will move the expression before the start of the loop and only execute it once. To keep the optimization from moving the concatenation expression on line 20 up in front of the for loop statement as a loop invariant, the test bench changes b each time around.

Example A-1. Simulation performance test bench

```
1    `ifdef SHORT
2    `define LIMIT 10
3    `else  // !SHORT
4    `define LIMIT 10_000_000
5    `endif // SHORT
6    module reverse;
7       integer i,j;
8       reg [0:7] e0,e1,b;
9    initial
10     begin
11       e0 = 8'h99;
12       e1 = 8'h99;
13       b = 8'd0;
14       for (i = 0; i < `LIMIT; i = i + 1) begin
15   `ifdef METHOD1
16          for (j = 0; j < 8; j = j + 1)
17             e0[j] = e0[j] ^ b[7-j];
18   `endif  // METHOD1
19   `ifdef METHOD2
20          e1 = e1 ^ {b[7],b[6],b[5],b[4],b[3],b[2],b[1],b[0]};
21   `endif  // METHOD2
22   `ifdef CHECK
23          if (e0 != e1) begin
24             $display(" e0 != e1 - %h !- %h",e0,e1);
25             $finish;
26          end
27   `endif  // CHECK
28          b = b + 8'd1;
29       end
30   `ifdef METHOD1
31       $display(" e0 = %h",e0);
32   `endif  // METHOD1
33   `ifdef METHOD2
34       $display(" e1 = %h",e1);
35   `endif  // METHOD2
36       $finish;
37     end
38   endmodule // reverse
```

B
Quick
Reference

This appendix describes the RTL Verilog language elements and organization designers supported for success in design verification.

Lexical elements

- <u>Names</u> - Begin with a-z A-Z or _, and may contain numbers 0-9. Examples: r_rreq, ck, rst_. *Names containing dollar signs are not supported.*
- <u>Unsized decimal integers</u> - Specify bit ranges, memory sizes, time. *High bit maximum: 255.* Examples: [7:0], [0:255], #1.
- **Sized integers** - use to represent bits. May be binary, octal, hex or decimal. *Maximum width: 256 bits.* Examples: 8'**b**0011_1001, 2'**o**2, 13'**h**03f, 9'**d**255
- Use "z" values only in binary to directly drive tri-state **output** or **inout** ports. *"x" values are not supported for verifiable RTL Verilog.*

Compiler Directives

- '**include** "*<file>*"
- '**define** *<name> <text to comment or end-of-line>*
- '**ifdef** *<name>*
- '**else**
- '**endif**
- '**timescale**

Reserved Words

bold - supported by verifiable RTL design in Verilog:

always	**else**	**initial**	**parameter**
assign	**end**	**inout**	**posedge**
begin	**endcase**	**input**	**reg**
case	**endfunction**	**module**	**tri**
casex	**endmodule**	**negedge**	**tri0**
default	**function**	**or**	**tri1**
defparam	**if**	**output**	**wire**

oblique - not supported by verifiable RTL design in Verilog:

and	*highz1*	*rcmos*	*task*
buf	*ifnone*	*real*	*time*
bufif0	*integer*	*realtime*	*tran*
bufif1	*join*	*release*	*tranif0*
casez	*large*	*repeat*	*tranif1*
cmos	*macromodule*	*rnmos*	*triand*
deassign	*medium*	*rpmos*	*trior*
disable	*nand*	*rtran*	*trireg*
edge	*nmos*	*rtranif0*	*vectored*
endprimitive	*nor*	*rtranif1*	*wait*
endspecify	*not*	*scalared*	*wand*
endtable	*notif0*	*small*	*weak0*
endtask	*notif1*	*specify*	*weak1*
event	*pmos*	*specparam*	*while*
for	*primitive*	*strong0*	*wor*
force	*pull0*	*strong1*	*xnor*
forever	*pull1*	*supply0*	*xor*
fork	*pulldown*	*supply1*	
highz0	*pullup*	*table*	

RTL Module

Module organization

module asic8 (*port declarations*);
    port directions
    port types ("reg" for non-tri-state outputs)
    global reg declarations
function declarations
always *@(sensitivity_list)*
  begin
    procedural statements (combinational logic)
  end
assign *statements (combinational logic)*
tri-state port driving **tri** *statements*
storage element module instances
endmodule

Port declarations

module asic8 (ck, scan_in, scan_ctl, bdp, scan_out);
  input ck;
  input scan_in;
  input [*1:0*] scan_ctl;
  inout [*17:0*] bdp;
  output scan_out;
  reg scan_out;   *// declare all non tri-state outputs "*reg*"*
                 // if driven by procedural assignments

Reg declarations

  reg c_b0_bnk0_busy; *// one bit variable*
  reg [7:0] r_p0_l0_zone; *// 8-bit variable*
  reg [1:0] M_valid [0:4095]; *// 4096x2 memory*

Functions

function [*<bit_range>*] *function_name*;
  inputs
  [local reg declarations]
begin
  procedural statements
end
endfunction   *// function_name*

Function example:

```
function csa_s1;
    input x;
    input y;
    input z;
  begin
   csa_s1 = expression;
  end
endfunction      // csa_s1
```

Tristate port driving statements

- Use to drive all **inout** and tri-state **output** type ports.
- Set port name equal to a single term. Put logic expression driving that term back in eval or eval_out task. Example:

    // inside procedural statement block
    c_bdp = (~csa_ & ~wa) ? r_a_reg : 18'bz;
     ...
    tri [0:17] bdp = c_bdp; *// preceding flip-flop macro calls*

Storage element module instances

- Macro call organization:

    typename (bit_width, power, instname,outname,clock,innames);

- Example:

    DFFT_X (44, h, reg_addra, r_addr[43:0], , clock, c_addr[43:0]);

Procedural statements

A *procedural block* is a statement or series of statements enclosed within a
begin end;

Control

```
if (expression)
   statement or procedural block
else                  // optional else followed by:
   statement or procedural block

case (scan_ctl)
  2'd0 : statement or procedural block
  2'd1 : statement or procedural block
  default : statement or procedural block
endcase

c_q0_rwy_ctl = {i_scan, c_adv_q, c_q1_act, c_q0_act};
casex (c_q0_rwy_ctl)
  4'b1_?_??: c_q0_rwy = r_q1_rwy;
  4'b0_0_?0: c_q0_rwy = {c_rwy_qctl, c_par_data};
  4'b0_0_?1: c_q0_rwy = r_q0_rwy;
```

```
        4'b0_1_00: c_q0_rwy = {2'h0, c_rwy_qctl[8:2], c_par_data};
        4'b0_1_01: c_q0_rwy = {c_rwy_qctl, c_par_data};
        4'b0_1_1?: c_q0_rwy = r_q1_rwy;
    endcase
```

Assignments

- Non-blocking: targets must be type reg; use in module definitions for flip-flop assignments:
    `r_icq_0 <= c_icq_0;`
- Blocking: targets must be type reg; use in procedural blocks and functions for combinational variable and output assignments:
    `c_i0_btab_active = c_i0_btab_out[24];`
- Continuous assignments: targets must be type wire; outside procedural blocks:
    assign c_tab_reset_ctr = r_tab_reset_ctr + 5'**h01**;
- ***deassign*** *and assignments with timing controls are not supported for verifiable RTL Verilog.*

Operators

Binary operators

Arithmetic	+	**a + b**	**addition**
(2's complement)	-	**a - b**	**subtraction**
	%	**a % b**	**modulo**
Bit-Wise	&	**a & b**	**and**
	|	**a | b**	**or**
	^	**a ^ b**	**exclusive or**
	~^	**a ~^ b**	**exclusive nor**
Logical	&&	**a && b**	**and**
	||	**a || b**	**or**
Relational	==	**a == b**	**equality**
	!=	**a != b**	**inequality**
	>	**a > b**	**greater than**
	<	**a < b**	**less than**
	>=	**a >= b**	**greater or equal**
	<=	**a <= b**	**less than or equal**
Shift	<<	**a << b**	**logical shift left**
	>>	**a >> b**	**logical shift right**

Unary operators

~	~a	invert a
^	^a	parity of a
~^	~^a	not parity of a
!	! a	logical not
&	&a	unary and
l	l a	unary or
~&	~&a	unary nand
~l	~l a	unary nor

Miscellaneous operators

? :	a ? b : c	conditional expression
{,}	{a,b,c}	concatenation
{{}}	{a{b}}	replication (b must be 1-bit)

Unsupported operators

The following operators are not supported for Verilog tools.

-	-a	*unary minus*
*	*a * b*	*multiply*
/	*a / b*	*divide*
===	*a === b*	*equality (0/1/X/Z)*
!==	*a !== b*	*inequality (0/1/X/Z)*

Operator precedence

DON'T COUNT ON IT! Use parenthesis to force precedence. Verilog operator precedence is not 100% defined. In the absence of parenthesis, different tools may treat operator precedence differently.

System tasks and functions

$display("*text and format specs*",*signal,signal, ...*);
$finish;
$time   *// function*
$write ("*text, format specs*",*signal,signal, ...*);
- Supported format specs: %b, %d, %h, %o, %s, %t, \t, \n, \\, \", \\,
                          %%, %m.